Lecture Notes in Biomathematics

Managing Editor: S. Levin

W0080313

56

Herbert W. Hethcote
James A. Yorke

Gonorrhea
Transmission Dynamics
and Control

Springer-Verlag Berlin Heidelberg GmbH

Editorial Board

H. J. Bremermann J. D. Cowan W. Hirsch S. Karlin J. B. Keller
M. Kimura S. Levin (Managing Editor) R. C. Lewontin R. May J. D. Murray
G. F. Oster A. S. Perelson T. Poggio L. A. Segel

Authors

Herbert W. Hethcote
Department of Mathematics, University of Iowa
Iowa City, IA 52242, USA

James A. Yorke
Institute for Physical Science and Technology
and
Department of Mathematics, University of Maryland
College Park, MD 20742, USA

AMS Subject Classification (1980): 92 A 15

ISBN 978-3-540-13870-9   ISBN 978-3-662-07544-9 (eBook)
DOI 10.1007/978-3-662-07544-9

This work is subject to copyright. All rights are reserved, whether the whole or part of the material
is concerned, specifically those of translation, reprinting, re-use of illustrations, broadcasting,
reproduction by photocopying machine or similar means, and storage in data banks. Under
§ 54 of the German Copyright Law where copies are made for other than private use, a fee is
payable to "Verwertungsgesellschaft Wort", Munich.

© Springer-Verlag Berlin Heidelberg 1984
Originally published by Springer-Verlag Berlin Heidelberg New York Tokyo 1984

Foreword

The success of public health programs depends both upon the logical application of proven strategies and also upon a healthy understanding of what is unknown. The work depicted in this monograph provides an excellent example of displaying the known and unknown about the epidemiology of gonorrhea. The authors have modeled their data in a way that was extremely useful in formulating approaches to gonorrhea control at the national level. Their efforts are a good example of how mathematical modeling has more than just theoretical utility.

The authors' research has helped eliminate a number of misconceptions which we had about gonorrhea epidemiology. In large part because of this work, we now believe (1) that gonorrhea prevalence adjusts rapidly to both changes in sexual behavior and also activities of control programs, (2) that prevalence oscillates seasonally around an equilibrium state determined by the current social and medical conditions, and (3) that this equilibrium moves as epidemiological conditions change. These ideas are important in evaluating the effects of our programs and in formulating new approaches. The depiction of a highly sexually-active "core" population, which is highly infected and causes a large fraction of all new cases of gonorrhea, has been quite stimulating. This concept of the core, and the resultant emphasis on "efficient transmitters," was a major factor which influenced revision of national strategies to control gonorrhea. Academic experts and practical public health officials alike will find this monograph to be a very valuable example of the utility of modelling to influence disease control programs. Public policy decisions depend on accurate information processed through objective analytic minds; use of models—including the discipline of decision analysis—facilitates "scientific administration." With programs of national scope, such as STD control, application of macroscopic modeling has a broad influence and benefits large numbers of people.

Ending on a personal note, as past and present directors of the national STD control program, more than just the product of the modeling has proven beneficial. It was also the process of interacting with the authors which helped stimulate innovation and clarity of thought. We thank them not only for their work but for their insight.

Paul J. Wiesner, M.D.
Director
Chronic Diseases Division
Center for Environmental Health
Centers for Disease Control

Willard Cates, Jr., M.D., M.P.H.
Director
Division of Sexually Transmitted
 Diseases
Center for Prevention Services
Centers for Disease Control

TABLE OF CONTENTS

PREFACE

This monograph describes the results of our gonorrhea modeling project which began in 1973. This research project was funded by Centers for Disease Control Contracts 200-76-0613 and 200-79-0949, by National Institutes of Health Grant AI-13233, and by National Science Foundation Grant MCS81-0217. Although some results have been described in journal articles (Yorke, Hethcote and Nold, 1978; Hethcote, Yorke and Nold, 1982), the details and mathematical basis of many results have not appeared before or have appeared only in reports to CDC and NIH.

For most communicable diseases it is understood how an infective can transmit the infection by contacts with others and how a disease spreads through a chain of infections. Because of the numerous complex interactions in a population, it is difficult to comprehend the large scale dynamics of disease spread without the formal structure of a mathematical model. An epidemic model uses a microscopic description (the role of an infectious individual) to predict the macroscopic behavior of disease spread through a community. The purpose of mathematical models is to achieve a better understanding of how the biological and sociological mechanisms influence disease spread. Fixed parameters which occur in the models must have a well understood epidemiological interpretation such as a contact rate or a duration of infection.

Comparisons can lead to a better understanding of the process of disease spread. It may be possible to compare different diseases in the same population, the same disease in different populations or the same disease at different times. One way of making these comparisons is to formulate models for the various situations and then to compare the parameter values. We have made comparisons involving measles, rubella, mumps, chickenpox and poliomyelitis in several papers (London and Yorke, 1973; Yorke and London, 1973; Yorke, Nathanson, Pianigiani and Martin, 1979; Hethcote, 1983).

Although planned experiments can be used to obtain information in many sciences, experiments with infectious diseases in human populations are generally not possible for ethical and practical reasons. The only data usually available is from naturally occurring epidemics or from the natural incidence of endemic diseases; unfortunately, even these data are not complete since many cases are not reported. Hence the most basic facts of transmission may be in doubt. Since repeatable experiments and accurate data are usually not available in epidemiology, **mathematical models can be used to perform needed** theoretical **experiments** with different parameter values and different data sets.

In order to use epidemic models for a particular disease, the capabilities and limitations of the models must be realized. It is often not recognized that many important questions cannot be answered using a given class of models. **The most difficult problem for the modeler is finding the right combination** of available data, an interesting question and a mathematical model which can lead to the answer.

Since infectious disease models furnish a means of assessing quantitative conjectures and of evaluating control procedures, they can be the only practical approach to answering questions about which control procedure is the most effective. Quantitative predictions of communicable disease models are always subject to some uncertainty since the models are idealized and the parameter values can only be estimated. **Predictions of the relative merits of several control methods are often robust** in the sense that the same conclusions hold for a broad range of parameter values and a variety of models. Various control methods for gonorrhea are compared in Chapters 4, 5 and 6 of this monograph.

Although some of the results described in this monograph may be useful for other sexually transmitted diseases (STDs), we have focused our attention here on gonorrhea. The incidences of other STDs such as genital herpes, caused by herpes simplex virus, and nongonococcal urethritis, often caused by Chlamydia trachomatis, are increasing dramatically in North America and Europe. Because practical diagnostic tools, control methods and specific treatments are often lacking for these other STDs, their incidence is increasing faster than the incidence of gonorrhea. There has also been an increase in the sexual transmission of diseases with agents such as hepatitis B virus, cytomegalovirus and Group B streptococcus (NIAID, 1980). The incidence of syphilis has decreased dramatically so that it is much less than the incidence of gonorrhea. The models used here would generally not be suitable for syphilis since individuals with syphilis go through several stages.

We hope that the unified presentation in this monograph will be of use to epidemiologists, scientists, mathematicians and students interested in sexually transmitted diseases or in how mathematical models can contribute to the understanding of disease transmission and control. We also hope that this monograph will encourage other studies of specific diseases using mathematical models.

We thank the personnel of the Division of Venereal Disease Control, Center for Prevention Services, Centers for Disease Control, Atlanta, Georgia for valuable discussions and suggestions. This project would not have been possible without their willingness to consider new ideas, to discuss concepts and to supply data. During our visits there we have consulted with Drs. John P. Brennan, Joseph H. Blount, James W. Curran, William W. Darrow, Beth Goldman, Gavin Hart, Harold W. Jaffe, Robert E. Johnson, Oscar G. Jones, Robert J. Kingon, Mark A. Kramer, Franklin R. Miller, William C. Parra, Gladys H. Reynolds, Richard B. Rothenberg, Ronald K. St. John, and Akbar A. Zaidi. We especially appreciate the cooperation of Dr. Ralph H. Henderson, Director of the Division of Venereal Disease Control until 1976 and Dr. Paul J. Wiesner, Director until 1983.

CHAPTER 1

BACKGROUND AND BASIC CONCEPTS

It is obvious that the characteristics of a specific disease must be understood thoroughly in order to model that disease. Gonorrhea incidence, transmission, symptoms, complications and treatments are described in sections 1.1 and 1.2. Various current and potential control procedures for gonorrhea are described in section 1.3. Among the several basic types of models for infectious diseases described in section 1.4, it is apparent that one type is suitable for gonorrhea. A brief description of previous work on the mathematical modeling of gonorrhea is given. The concepts of the contact number and the infectee number are described in section 1.5.

1.1 The Magnitude of the Problem

As seen in figure 1.1, gonorrhea is the most frequently reported disease in the United States. Figure 1.2 shows that reported gonorrhea incidence in the United States tripled between 1965 and 1975. The increase of reported cases in this period is thought to represent an actual increase in gonorrhea incidence as well as improved procedures for case detection. Incidence is the number of new cases in a time interval such as a year, a month or a week. Reporting of gonorrhea, particularly by private health care providers, is incomplete in the United States so that reported incidence can only be used for determining trends or for comparing segments of the population.

Since 1975 the number of reported cases of gonorrhea has been about 1 million cases per year. Supplementary information such as extensive surveillance studies or polls of practicing physicians have been used to estimate that the **actual incidence of gonorrhea is now approximately two million cases per year in the United States.** The leveling off of reported cases after 1975 may be due to the gonorrhea control activities initiated in 1973 (CDC, 1981a).

Many countries have had rapid increases in the incidence of gonorrhea. The incidence of gonorrhea in several western countries including the United States, Canada, Denmark, Finland, Norway and United Kingdom was approximately level from 1955 to 1965 and generally increased from 1965 to 1975. Factors which may be responsible for the increased rates include: more frequent changes of sex partner, increased population mobility, increasing use of oral contraceptives, decreasing use of the condom and the diaphragm for contraception and

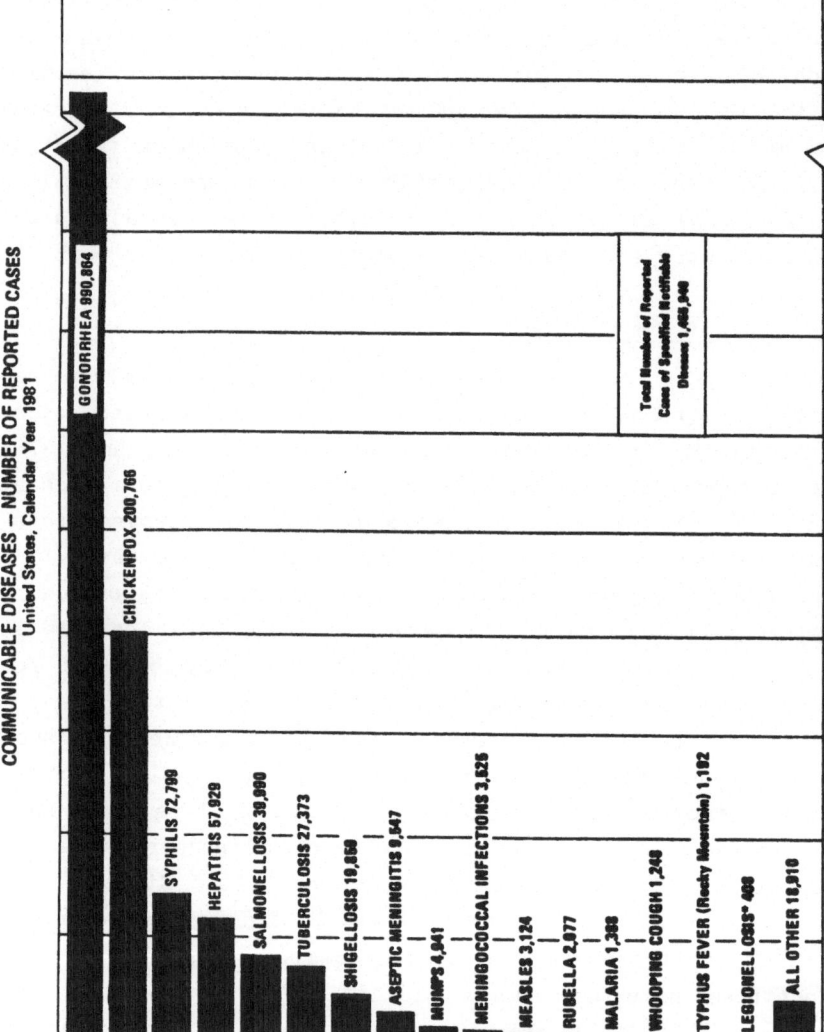

Figure 1.1. Reported cases of communicable diseases in the United States in 1981. Figure and permission from Statistical Services Section, Division of Venereal Disease Control, Centers for Disease Control.

GONORRHEA, UNITED STATES, CALENDAR YEARS 1950-1982

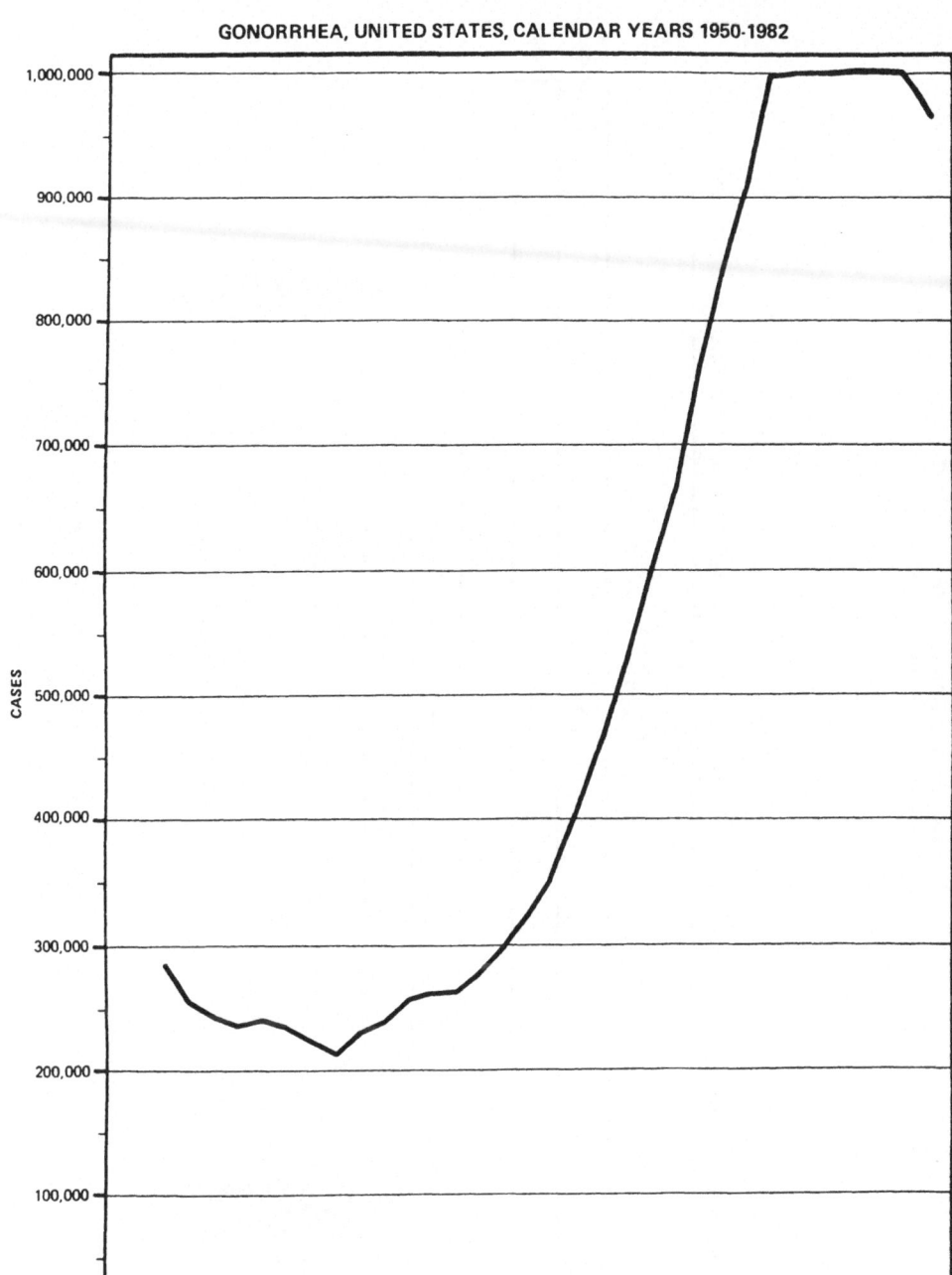

Source: CDC 73.688, HHS,PHS,CDC,CPS,DVDC,SSS, Atlanta, Georgia 30333

Figure 1.2. Reported cases of gonorrhea in the United States from 1950
to 1982. Figure and permission from Statistical Services Section,
Division of Venereal Disease Control, Centers for Disease Control.

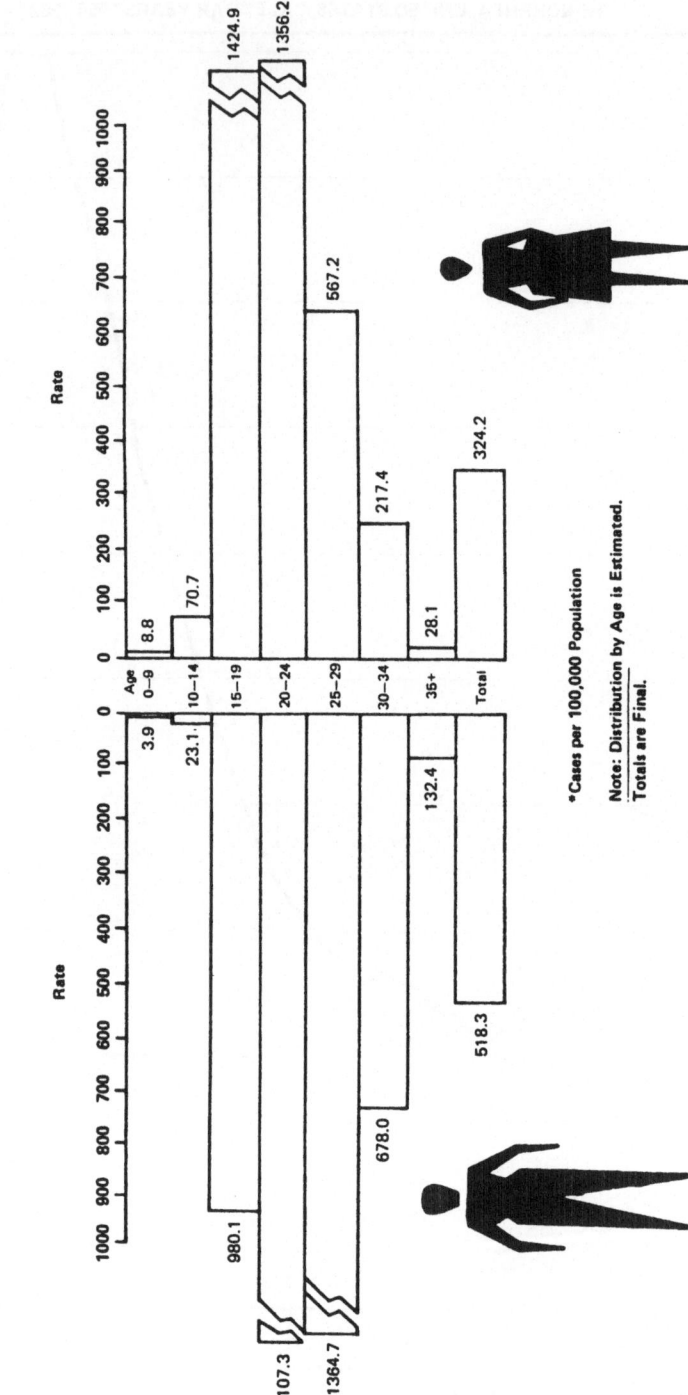

Figure 1.3. Sex and age-specific case rates of gonorrhea in the United States in 1982. Figure and permission from Statistical Services Section, Division of Venereal Disease Control, Centers for Disease Control.

increasing gonococcal resistance to antibiotics (WHO, 1978). Reporting is relatively accurate in the United Kingdom where most cases are seen in clinics and in Denmark where most cultures are processed in a single laboratory. In the developing countries of Asia, Africa and Latin America, the reporting of gonorrhea is often very incomplete, but **annual incidences as high as 26% have been reported**. In those countries where extramarital sexual intercourse is common for men but not for women, a high proportion of men with gonorrhea are infected by prostitutes (WHO, 1978).

The demographic factors that correlate best with gonorrhea incidence are age, race, marital status, socioeconomic status and urban residence. Individuals who are single, have a lower socioeconomic status and reside in a large city are more likely to be infected by gonorrhea. As seen in figure 1.3 the age-specific incidence rates in the United States are highest for young adults (20 to 24 years of age) and second highest for teenagers (15 to 19 years of age). In 1978 young adults (ages 20-24) accounted for 39 percent and teenagers for 25 percent of reported gonorrhea cases. **Approximately one of 30 teenagers will acquire gonorrhea this year** (NIAID, 1980).

The great concern about the high incidence of gonorrhea results from the severe complications that some women suffer. Pelvic Inflamatory Disease (PID) is the most important complication. About 10 to 20 percent of women with gonococcal infection will suffer from PID. Data from the National Center for Health Statistics show that approximately **1 million women are treated for PID each year** including 212,000 who are hospitalized (about 50,000 required abdominal surgery). The direct hospitalization cost of PID in the United States has been estimated to be more than $600 million annually (Curran, 1980; NIAID, 1980). Infertility, ectopic pregnancy and chronic pelvic pain are important delayed consequences of gonococcal infection. In 1976 approximately 80,000 women were made sterile by gonococcal PID. The incidence of ectopic pregnancy rose from about 13,200 cases in 1967 to more than 41,000 cases in 1977 (NIAID, 1980; CDC, 1980b).

PID occurs when the fallopian tubes become swollen and inflamed due to infection. In many women who are unaware that they are infected with gonorrhea, the gonococci rise from the vagina and cervix into the uterus and then during menstruation spread up the sides of the uterus and into the fallopian tubes. When the inflamed fallopian tubes heal, the scarring may block the fallopian tubes causing involuntary fertility. Women with one episode of gonococcal PID have a 6% chance of becoming infertile and the likelihood of infertility

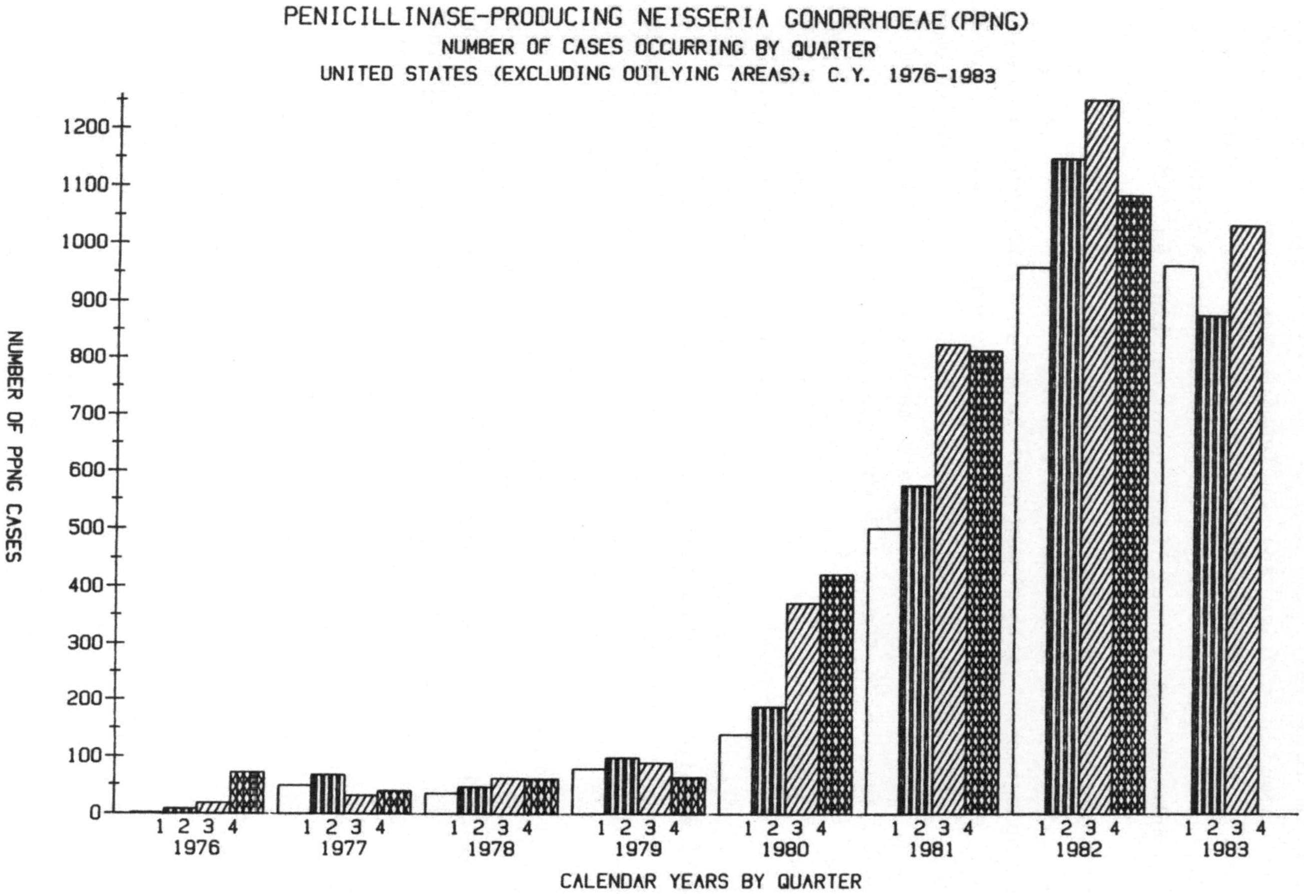

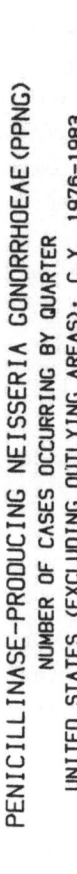

Figure 1.4. Reported cases of PPNG by quarter in the United States. Figure and permission from Statistical Services Section, Division of Venereal Disease Control, Centers for Disease Control.

increases with each subsequent episode. Since partial obstruction of the fallopian tubes can lead to ectopic pregnancy, women who have had PID are 6 to 10 times as likely to have an ectopic pregnancy as those who have not had PID. Adhesions on the ovaries, bowel or other tender pelvic structures that occur during the healing of acute PID cause chronic pelvic pain in 10% to 20% of patients (NIAID, 1980; Rein, 1977; WHO, 1978).

Some women are more likely to get PID than others. Women who use intrauterine devices (IUDs) are estimated to be about 3 times as likely to get PID from a gonococcal infection as those who use no contraceptive measures. Women who have multiple sex partners and women who have had PID before are more likely to get PID. Women who use oral contraceptives are less likely to get PID. Women who use barrier methods of contraception such as diaphragms, condoms, spermicidal foams, jellies or creams, are also somewhat protected against PID (WHO, 1978).

Penicillinase-producing Neisseria Gonorrhoeae (PPNG) consists of strains of gonococci which are resistant to all forms of penicillin. Penicillinase is an enzyme that destroys penicillin. **PPNG has now spread throughout the world** and 2 to 6 fold increases in PPNG have occurred in some countries in the last few years (CDC, 1982). Despite several early successful efforts in eliminating PPNG when it was introduced in the United States, it now appears to have become permanently established. Figure 1.4 shows that the number of reported cases of PPNG infection in the United States has increased every year since 1976. Not only are there more cases of PPNG infection imported from Southeast Asia, but also there is now sustained domestic transmission in major cities such as Los Angeles, New York City and Miami (CDC, 1982).

The standard treatment for gonococcal infection is the administration of penicillin. Because more expensive antibiotics are needed to cure PPNG, the treatment of gonorrhea could rapidly become more difficult and generally less effective so that incidence rates could increase. Efforts to control PPNG transmission in the United States include testing some gonococcal isolates for PPNG, rapid identification and tracing of sexual partners of PPNG patients, screening of groups who are at high risk of PPNG infection, and treatment of the following with spectinomycin: PPNG patients and their sexual partners, patients who were infected in countries with high PPNG prevalence, and patients for whom penicillin, ampicillin, amoxicillin or tetracycline was not effective treatment for gonorrhea (CDC, 1980).

In areas where more than 5% of gonococcal isolates are PPNG, it is recommended that the more expensive drug spectinomycin be used as the initial treatment for all cases (CDC, 1980a). Selection for drug resistant strains continues; recently, spectinomycin-resistant PPNG was found (CDC, 1981b).

1.2 The Infected Person

The gonococcus bacteria that causes gonorrhea grows well only on mucous membranes and dies in seconds outside the human body. The probability of transmission of gonococcal infection during a single sexual exposure from an infectious woman to a susceptible man is estimated to be from .2 to .3, while the probability of transmission from an infectious man is approximately .5 to .7 (Wiesner and Thompson, 1980). If sexual intercourse occurs with an infected partner several times, the probability of transmission of infection is increased. Taking these repetitions into account, it seems that in casual liasons an infectious man is roughly twice as likely to infect a susceptible woman as when the roles are reversed.

Gonococcal infection does not seem to confer immunity; more precisely, **none of the immunologic defense mechanisms has been shown to consistently prevent reinfections** (NIAID, 1980). This may be due to the great antigenic diversity of gonococcal surface antigens. In order for an individual to become infected, the gonococci must attach themselves to mucosal surfaces so that they will not be washed away by urine and mucosal flows. The gonococci have hair-like appendages called pili on their surfaces which facilitate their attachment. Local antibodies to Neisseria gonorrhoeae have been found on the mucosal surfaces which inhibit attachment of gonococci to epithelial cells; however, these antibodies disappear after the gonococcal infection ends. Humoral antibodies to N. gonorrhoeae have been demonstrated by a number of assays, but the presence of these antibodies has not been shown to correlate in any way with protection from reinfection. Lymphocyte activation and production of migration inhibition factor to various gonococcal antigens has been shown; however, the presence of a cellular immune response does not protect against reinfection. In individuals with Pelvic Inflamatory Disease the presence of bactericidal antibody does not protect individuals from either subsequent reinfection or PID. Although gonococcal infections usually remain localized on mucosal surfaces, some strains of gonococci can cause disseminated gonococcal infection. These strains seem to have a surface antigen which binds an IgG antibody capable of blocking killing of the gonococci by normal human serum. Even disseminated gonococcal infection does not lead to humoral or cellular immunity that protects against reinfection (NIAID, 1980).

In men, the initial infection is in the anterior urethra so that some of the infecting organisms are removed with each urination Approximately 90 percent of all men who have a gonococcal infection notice symptoms within a few days after exposure and promptly seek medical treatment. Another 5 percent have mild or transient symptoms and the remaining 5 percent have totally asymptomatic infections. These latter two groups of men often fail to seek treatment and, consequently, are responsible for much of the transmission of disease to women. As many as half of the women with complications of gonorrhea were infected by asymptomatic men (NIAID, 1980). It has been estimated that approximately 5% of the incidence in men are asymptomatic and **asymptomatic men account for 60-80% of the prevalence** and hence 60-80% of the transmissions (Wiesner and Thompson, 1980). Complications of gonorrhea in men are rare today.

Studies of the course of the infection in women have been found to be dangerous since an infected woman is likely to develop PID during such a study. Consequently, investigations conducted before penicillin became available are often the most definitive studies available. One 1942 study analyzing prolonged untreated gonococcal infections among 73 inmates of the New York City House of Detention for Women is summarized below (Mahoney et al., 1942).

Most of the women were prostitutes. Each woman initially had a positive culture test for gonococci, and each was symptomatic. Each patient was cultured two or three times a week, when possible, for three to four months. Ten to twenty culture plates were used on each exam for each patient, instead of the customary two. Three patterns were observed. In each type the clinical symptoms usually remained without appreciable alteration throughout the observation period. No treatment was administered during the study; however, at least some patients were treated after the study before being released.

Type 1 (46% of the patients) remained positive. Usually each plate at an examination yielded approximately equal numbers of colonies.

Type 2 (42%) became and remained culture negative, though symptomatic.

Type 3 (12%) reverted to negative and after numerous consecutive negative findings had one or more positive findings sporadically appearing.

Due to the continued symptoms of all types, the investigators felt strongly that gonococcal infection remained despite the negative cultures. Menstrual periods had a negligible effect on the positivity or negativity of the examinations.

Although some reports indicated that the use of oral contraceptives affected the susceptibility of women, studies in venereal disease and family planning clinics show that the prevalence of gonococcal infections is not increased in women using oral contraceptives. However, hormonal factors do influence the clinical appearance of gonorrhea so that half of women with gonococcal PID seek medical attention in the first week of the menstrual cycle (WHO, 1978).

One recent estimate is that 50 to 75 percent of women with gonorrhea fail to have symptoms which cause them to seek medical care (NIAID, 1980). Another estimate is that 30-60% of the incidence in women is asymptomatic and that **asymptomatic women account for 80-90% of the prevalence** in women (Wiesner and Thompson, 1980). Asymptomatic women often remain untreated either until they develop PID or disseminated gonococcal infection or until they are examined and treated because they are suspected of transmitting infection or until they are discovered through mass screening by culture testing. Asymptomatic women and men form a vast reservoir of persons infected with gonorrhea. Transmission between homosexual men produces another reservoir of anorectal and/or pharyngeal infection (NIAID, 1980).

Precise diagnosis in both men and women depends on laboratory testing. The standard laboratory test for gonococci consists of bacteriological culturing of the patient's secretions. Serological tests are no longer used because of their low sensitivity and specificity (WHO, 1978). Except for PPNG, a single dose of penicillin is very effective in curing gonococcal infection. Other more expensive antibiotics are effective for PPNG.

1.3 Control Procedures

Gonorrhea control procedures are constantly being reevaluated in order to achieve the optimal use of available resources. One primary goal of this monograph is to compare gonorrhea prevention activities by means of mathematical models.

Educational programs either in clinics or in the media might make the sexually active population more aware of the symptoms and seriousness of gonorrhea so that people who suspect that they might be infected would seek examination and treatment sooner. Educational programs can promote post exposure prophylaxis methods such as urination, local genital cleansing, douching and local antisepsis after intercourse. Individuals who use condoms are less likely to get gonorrhea. A national campaign in Sweden promoting condoms may have been a factor in the reduction of gonorrhea incidence there (WHO, 1978).

Screening. Before 1972 nearly all sexually transmitted disease (STD) control activities at the state and local level in the United States were syphilis case finding and prevention efforts. The major dangers of gonorrhea were not generally recognized before 1972. Gonorrhea control consisted of diagnosis and treatment of individuals who came to public clinics. By 1974 nearly all state and local health departments had established federally-assisted gonorrhea control programs. A primary part of these programs involves culture testing large numbers of women for gonorrhea. Although some of the culture tests are for patients with symptoms who have come to clinics, many of the tests are for women without symptoms who are having a gynecologic examination. Adding a gonorrhea culture test to a pelvic examination is relatively inexpensive since the federal government pays for the laboratory analysis. The test is administered to women who are sexually active and in age groups where gonorrhea is present. **Each year since 1974 there have been over 8 million gonorrhea culture tests on women** of which between 4 and 5% were positive for gonococcal infection (CDC, 1979a). A goal of this extensive screening program is the identification and treatment of women with asymptomatic disease.

Some of the positive culture tests were diagnostic in the sense that they verified infection in women who had symptoms suggesting gonococcal infection or who had sexual contacts with infected individuals. Some of the positive culture tests were discoveries of the screening program in the sense that they occurred when there was no reason to expect infection or as the result of a routine pelvic examination. Yorke, Hethcote and Nold (1978) have estimated the effectiveness of the screening program in the United States in discovering infectives by using data from 1967-75. It was estimated that in 1973-75 approximately a third of the reported cases of gonorrhea in women were discoveries of the screening program and that approximately a tenth of all actual cases of gonococcal infections in women were discovered by the screening program. Reported case rates in men are largely independent of public awareness or screening campaigns. By assuming that the reported incidence in men is proportional to actual incidence in men, it was estimated that the 1974 screening program for women caused the **actual male incidence** to decrease so that it **was approximately 20% below what it would have been without screening.** In Chapter 6, we use the estimates above to choose a parameter value. Specifically, we adjust a parameter value so that discovering (and treating) 10% of the infected women via the screening program yields a 20% decrease in male incidence.

Contact Tracing and Interviewing. Contact investigation or tracing attempts to identify contacts of known infectives and to encourage contacts to be checked as soon as possible. Intensive contact investigation is a very important control method for syphilis. Contact investigation for gonorrhea sometimes consists of educating known infectives about the seriousness of gonorrhea and asking them to encourage their contacts to be examined. Some clinics try to obtain identifications of known contacts, to communicate with the contacts by phone or in person and to encourage them to be examined for gonococcal infection. Early identification of infectives by contact tracing reduces their infectious periods and, consequently, can reduce the chance of transmitting the infection. During 1982 about 360,000 patients with gonorrhea were interviewed for sex partner referral, about 335,000 contacts were obtained, about 215,000 partners were examined and about 172,000 were treated either therapeutically or preventively (CDC, 1983). Many additional patients were counseled about their infection and advised to refer their sex partners for examination.

Vaccines. Although **no vaccine which prevents gonorrhea is now available,** many researchers are trying to develop a practical gonorrhea vaccine (WHO, 1978; NIAID, 1980; Marx, 1980; Shearer, 1983). One vaccine containing gonococcal pili elicited the production of antibodies both in blood and in the secretions of the urogenital tract, where they may prevent bacterial attachment (Marx, 1980). Another gonococcal pili vaccine stimulated the production of antibodies in the blood, but had only a slight, temporary effect on the secretory antibodies of the urogenital tract. Since there are about 1000 identifiable strains of gonococci with different pili, a gonorrhea vaccine would have to contain a variety of pili. The three exterior cellular components of a gonococcus are pili, proteins of the outer membrane and lipopolysaccharides so that some scientists suggest that a gonorrhea vaccine could be based on the latter two components (WHO, 1980). Because both gonorrhea and meningococcal meningitis are infections of mucosal surfaces and because the recently developed vaccine for meningitis only provides immunity for several months, it is possible that a gonorrhea vaccine would only provide temporary immunity. Vaccines that produce permanent immunity are generally directed at viral diseases.

The search for a safe and effective gonorrhea vaccine continues. A recent news article (Shearer, 1983) indicates that the U.S. Army has field-tested an experimental vaccine for gonorrhea in Korea. Half of the 5000 volunteers in the study were injected with two doses of the Gonococcal Pilus Vaccine and the other half were given two doses of a placebo. The usefulness of this vaccine is unknown since the results of the study have not been announced.

1.4 Modeling Transmission Dynamics

In describing the transmission dynamics of a communicable disease, it is convenient to divide the population into disjoint classes whose sizes may change with time. The susceptible class S contains those can become infected, the exposed class E contains those who are in a latent period but are not yet infectious, the infective class I contains those who are infectious, and the removed class R contains those who have at least temporary immunity either from immunization or previous exposure.

There are several basic types of epidemiologic models (Hethcote, 1976; Hethcote, Stech and van den Driessche, 1981c). To model an epidemic (i.e., a sudden unusual increase in cases) of a disease for which recovery confers permanent immunity, SIR or SEIR models without vital dynamics (births and deaths) are appropriate. A sequence of letters such as SEIR describes the movement of individuals between the classes: susceptibles become latent, then infectious and finally recover with immunity. To model diseases which confer permanent immunity and which are endemic (i.e., always present) because of births of new susceptibles, SIR or SEIR models with vital dynamics are suitable. Models of SIRS or SEIRS type are used to model diseases which remain endemic because the immunity is only temporary. Diseases for which recovery does not confer immunity are described by SIS or SEIS models.

Gonorrhea has three striking epidemiological characteristics which must be incorporated into a model. First, as described in section 1.2, gonococcal infection does not confer protective immunity so that individuals are susceptible again as soon as they recover from infection. Indeed, this lack of protective immunity makes gonorrhea very different from other diseases such as measles, mumps, rubella, chickenpox, poliomyelitis, diphtheria, whooping cough, and tetanus.

Gonorrhea is a particularly interesting disease to model precisely because infection does not confer protective immunity. **Second, individuals who acquire gonorrhea become infectious within a day or two** so that the latent period is very short compared to the latent period of about 12 days for measles, 15 days for chickenpox, and 18 days for mumps. Indeed, the latent period for gonorrhea is short enough so that it is not necessary to include an exposed class in a model for gonorrhea. **Third, the seasonal oscillations in gonorrhea incidence are very small** (less than 10%). In contrast, the incidences of diseases such as influenza, measles, mumps and chickenpox often vary seasonally by factors of 5 to 50 or more. Thus, models which use constant values for parameters such as the contact rates provide good approximations for gonorrhea.

Because the three characteristics of gonorrhea described in the previous paragraph seem to imply that SIS models with constant parameter values are suitable for modeling gonorrhea, we now investigate this issue further. Time delays can be introduced to model the time required for a person to pass from one disease state to another. Since periodic solutions can arise in SIRS models with time delays even though the parameter values are constant (Hethcote, Stech and van den Driessche, 1981a), one might wonder if periodic solutions can arise in SEIS models with time delays and constant parameter values since SEIS models are also cyclic with 3 classes; however, it has been shown that these SEIS models have stable equilibrium points so that they do not have periodic solutions arising by Hopf bifurcation (Hethcote, Stech and van den Driessche, 1981b). Indeed, for both SEIS and SIS models with constant parameter values, time delays do not change the general nature of thresholds or asymptotic stability; in all constant parameter models (i.e., without seasonal variation) the disease either dies out or approaches an endemic steady state (Cooke and Yorke, 1973; Hethcote, Stech and van den Driessche, 1981c). Because the essential behavior of SEIS models is the same as for SIS models, SIS models with no latent period are accurate approximations to SEIS models with very short latent periods. Since time delays do not affect the asymptotic behavior and since we will be interested in the behavior near the equilibrium points or in how the equilibrium points change when parameter values or control procedures change, the ordinary differential equation models without time delays are sufficiently general (Hethcote and Tudor, 1980). In conclusion, **SIS models** using ordinary differential equations without seasonal variation and without time delays **are satisfactory for describing the transmission**

dynamics of gonorrhea.

The first mathematical model explicitly for gonorrhea was developed by Cooke and Yorke (1973). They studied the asymptotic behavior of solutions of an integral equation model for a single homogeneous population which used time delays to represent variation in the infectious period. Cooke (1976) considered the effects of immigration on this model. Reynolds and Chan (1974) considered a linear differential equation model for gonorrhea, estimated the parameters, and projected the prevalence for women and men, both with and without terms modeling control procedures. Because the model is linear, the prevalence can grow exponentially without saturating the populations. Constable (1975) discussed the problems of gonorrhea modeling and formulated a model with five groups for each sex. Hethcote (1973) showed that all endemic solutions of an ordinary differential equation SIS model with a periodic contact rate approach an explicitly given periodic solution. A two group differential equation model for gonorrhea has been analyzed by Hethcote (1974,1975,1976). Bailey (1975, Chapter 11) showed how SIS models developed for other diseases can be used for gonorrhea.

A model for gonorrhea with an arbitrary number of interacting groups was formulated and analyzed by Lajmanovich and Yorke (1976) and Nold (1980). The section on gonorrhea in the differential equations text by Braun (1975) is based on material (used without acknowledgment) from a preprint of the paper of Lajmanovich and Yorke (1976). Aronsson and Mellander (1980) showed that if the Lajmanovich and Yorke model is modified to include seasonal variation in the contact and removal rates, then in the endemic case, there is a nontrivial periodic solution which is globally asymptotically stable. Nallaswamy and Shukla (1982) modified the Lajmanovich and Yorke model to include spatial diffusion and analyzed the stability of the endemic equilibrium state. Thieme (1982) showed that the Lajmanovich and Yorke global stability results still hold if short periods of incubation or immunity are included. Hirsch (1984) obtained the Lajmanovich and Yorke global stability results for a more general model.

Yorke, Hethcote and Nold (1978) argued that gonorrhea prevalence responds rapidly to changes in social behavior and control procedures. They also showed that the equilibrium prevalence moves as social and medical conditions change. They introduced concepts such as saturation, preemption and the core subpopulation into gonorrhea analysis. These results are described in more detail and justified mathematically in this monograph. Kemper (1978,1980) studied SIS models for one population with asymptomatics (infectives without symptoms) and with superspreaders (highly infectious individuals). Bailey (1979) reviewed the current state of modeling gonorrhea and formulated some new models with symptomatics and asymptomatics. The local stability of these models was analyzed by Wichmann (1979).

Kramer and Reynolds (1981) used a stochastic computer simulation model to evaluate gonorrhea vaccination and other control strategies. Hethcote, Yorke and Nold (1982) used an eight group model for gonorrhea to compare the effectiveness of six prevention methods for gonorrhea involving population screening and contact tracing of selected groups. The population was divided according to sex, sexual activity and symptomatic or asymptomatic infection. These results are presented and explained in Chapter 6. A news article about the results in this paper of Hethcote, Yorke and Nold (1982) appeared in Nature (May, 1981). Cooke (1982) proved global asymptotic stability for an SIS model involving symptomatics and asymptomatics. See Cooke (1984) for a survey of infectious disease models with asymptomatics.

1.5 The Contact Number and the Infectee Number

The contact number for an infectious disease in a population is defined as the average number of adequate contacts of a typical infectious person during the infectious period (Hethcote, 1976). An adequate contact is a direct or indirect contact that is sufficient for transmission of infection if the individual contacted is susceptible. The concept of an adequate contact is necessary since transmission of infection sometimes does not occur during sexual intercourse between an infective and a susceptible. Moreover, the probability of transmission from an infectious man to a susceptible woman is greater than the probability when the roles are reversed. If all people contacted were susceptible, then the contact number would be the average number of people infected by one infective during the infectious period. The contact number for gonorrhea depends on the sexual behavior of the population being considered. The contact number has also been called the reproduction rate (Dietz, 1975).

Some of the adequate contacts of infectious individuals do not result in transmission of the infection since these contacts are not with susceptibles. An infectee is someone infected by an adequate contact. The infectee number for an infectious disease is the average number of actual transmissions by an average infective during the infectious period. The infectee number may vary with time, for example, seasonally. The infectee number is the product of the contact number and the susceptible fraction at the given time. The infectee number has also been called the replacement number (Hethcote, 1976), the infector number (Yorke, Hethcote and Nold, 1976) and the reproduction rate (Anderson and May, 1982).

A basic principle for all diseases is that **the infectee number is one when the disease is at an endemic equilibrium** (i.e., a positive steady state) (Hethcote, 1976; Yorke, Hethcote and Nold, 1978; Nold, 1979). This principle is an average result since some infectives might infect several susceptibles and other infectives might infect no susceptibles. It is assumed that each transmission is the result of a person to person contact (i.e. there are no vectors) and that the population is closed (i.e., there is no immigration of infected people). The principle above is intuitively reasonable since if the average infective passed the disease on to more than one susceptible person, then the prevalence would increase. If the average infective passed the disease on to less than one individual, the prevalence would decrease.

One consequence of the principle above is that at an endemic equilibrium the average infective has two adequate contacts during the course of the infection: a contact with an infector or source of the infection and a contact with an infectee to whom the infection is transmitted. Thus the average person with gonococcal infection would have at least two sex partners during the period of infection. See Yorke, Hethcote and Nold (1978) for a comparison of this theoretical result with some clinical survey data.

CHAPTER 2

A SIMPLE MODEL FOR GONORRHEA DYNAMICS

The SIS model in section 2.1 where susceptibles become infectious and then susceptible again is based on the careful description in Chapter 1 of the characteristics of gonorrhea: **there is negligible protective immunity, negligible latent period and negligible seasonal oscillations.** It is the simplest possible model since it assumes that gonorrhea transmission occurs in one uniform, homogeneous population. The population represented by the model would necessarily consist only of those individuals at high risk who are also efficient transmitters. Thus people who are less active sexually would not be represented in this model. While the model restricts attention to this group, the model does not indicate the size of this group. Notice that this model ignores the epidemiological differences between women and men.

This model introduces notation and, like the more refined models later in this monograph, it has a threshold which determines whether the disease dies out or approaches an endemic equilibrium point. The incidence at the endemic equilibrium in the model depends on specific parameter values and this equilibrium will move as these parameter values change. The concept in section 2.2 of **a moving equilibrium provides a basis for understanding observed changes and for predicting changes in incidence** resulting from changes in epidemiological factors. In section 2.3 the **rapid response of gonorrhea incidence to epidemiologic changes is justified by both observations and calculations.**

The SIS model considered here gives us a theoretical framework to use in drawing simple conclusions. We can now see the fallacies in a variety of ideas that were widely held. In the seventies some observers thought that the increase in reported gonorrhea incidence was similar to exponential growth and that it would continue to increase exponentially. For example, a Scientific American (1976) news article stated, "Today gonorrhea is an epidemic disease out of control. . . . Reversing the exponential increase in gonorrhea calls for a two-pronged attack: . . ." A few people thought that the gonorrhea epidemic would follow a classic epidemic curve as observed for an SIR model without vital dynamics; thus they expected incidence to rise to a peak and then decrease. The gonorrhea model presented in this chapter and the concepts of a moving equilibrium and rapid response which follow from the model provide a careful analysis and correct the misconceptions above.

2.1 One Population Model for Gonorrhea

Assume that the population considered has a constant size N which is sufficiently large that the sizes of each class can be considered as continuous variables instead of discrete variables. The fractions of the population that are susceptible and infectious at time t are $S(t)$ and $I(t)$, respectively. The fraction $I(t)$ is called the prevalence. As noted in section 1.4 the exposed class of latent individuals is ignored since the latent period is very short. There is no acquired protective immunity.

The contact rate λ is the average number of adequate contacts of an infective per day. An adequate contact is a direct contact during sexual intercourse which is sufficient for transmission of infection if the individual contacted is susceptible. Thus the average number of susceptibles infected per day by the infective class of size NI is λSNI. Here the contact rate λ is assumed to be fixed and does not vary seasonally. We remark that the population is uniform and homogeneously infected in the sense that each person having an adequate contact has the same probability of contacting an infective (namely, the probability $I(t)$).

Here we let d be the average infectious period and assume that the average infective has a 1/d chance of recovering on any day, independent of how long the person has been infected. This assumption is equivalent to the assumption that individuals recover and become susceptible again at a rate proportional to the number of infectives NI with proportionality constant 1/d. It is also equivalent to the assumption that the duration of infection has a negative exponential distribution (Hethcote, Stech and van den Driessche, 1981c).

Since $S(t) = 1 - I(t)$, the initial value problem for the number of infectives is

$$\frac{d}{dt}(NI(t)) = \lambda NI(t)(1-I(t)) - NI(t)/d , \quad NI(0) = NI_o \qquad [2.1]$$

After division by N, the differential equation and initial condition for the prevalence $I(t)$ become

$$\frac{dI}{dt} = \lambda I(1-I) - I/d , \quad I(0) = I_o \qquad [2.2]$$

For this model the contact number σ defined in section 1.5 is equal to the product of the daily contact rate λ and the average infectious period d in days.   Figure 2.1 shows the susceptible and infective compartments and the transfer rates between compartments.

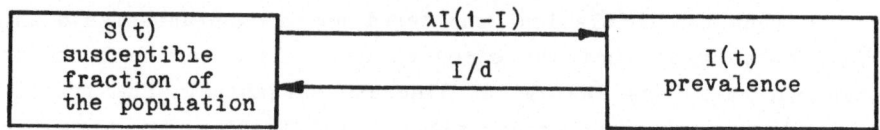

Figure 2.1   Flow diagram for the model [2.1].

The solution of [2.2] has the explicit form

$$
I(t) = \begin{cases} \dfrac{e^{(\sigma-1)t/d}}{\sigma(e^{(\sigma-1)t/d}-1)/(\sigma-1) + 1/I_o} & \sigma \neq 1 \\[4mm] \dfrac{1}{\lambda t + 1/I_o} & \sigma = 1 \end{cases} \qquad [2.3]
$$

The asymptotic results below follow from this solution.

If the contact number satisfies $\sigma < 1$ and initially there is at least one infective, then the infectee number $\sigma S(t)$ always satisfies $\sigma S(t) < 1$ so that the average infective is replaced by less than one infective.   Thus the disease will eventually die out (i.e., $I(t) \rightarrow 0$ as $t \rightarrow \infty$) since the average infective is not being replaced by at least one new infective.

If the contact number satisfies $\sigma > 1$, then the average infective can be replaced if the susceptible fraction of the population is high.   In this case, the disease remains endemic and the prevalence $I(t)$ approaches the positive equilibrium or steady state value $1-1/\sigma$ as t approaches ∞ .   At the endemic equilibrium point, the susceptible fraction of the population S is $1/\sigma$ so that the infectee number satisfies $\sigma S = 1$ as predicted in section 1.5.   In summary we point out that the **contact number 1 is the threshold which determines   whether   the   disease   dies   out ($\sigma<1$) or   remains endemic ($\sigma>1$)** .

The contact number depends on both the disease and the population being considered.   A contact number may be greater than 1 in one population and less than 1 in another.   For example, the male gay community has a large number of diseases that do not usually propagate heterosexually: rectal warts, hepatitis B and AIDS (Acquired Immuno-deficiency Syndrome).

In Chapter 1 we defined incidence as the number of new infectives per unit time. The daily incidence in our model [2.1] is $\lambda NI(1-I)$. Since we deal with fractions of the population in our mathematical models such as [2.2], we have defined prevalence as the fraction $I(t)$ of the population that is infectious at a given time as opposed to the number $NI(t)$ of people in the population who are infectious. For this and subsequent models, **when the disease is at an equilibrium, the prevalence times the population size is equal to the incidence times the duration.** Here we see this from [2.1] since the derivative of $NI(t)$ is zero at an equilibrium point. We remark that epidemiologists usually define prevalence as the number of infectious individuals at a given time so that their relationship is that prevalence equals incidence times duration.

2.2 Changes in Incidence: A Moving Equilibrium

The epidemiologic factors of a disease are the characteristics of the disease and its environment that affect transmission. **Epidemiologic factors include sociological aspects** such as contact rates among individuals and among subpopulations, sizes of the affected population and subpopulations, social and economic conditions, psychological attitudes and control programs. **They also include clinical aspects** such as average durations of the incubation, latent and infectious periods, virulence of the agents and their resistance to certain treatments, and availability and quality of medical care. The epidemiologic factors at a given time determine a theoretical equilibrium or steady state level and if the epidemiologic factors do not change, the actual incidence will approach this equilibrium level. In the model in section 2.1, the prevalence $I(t)$ approaches $(1-1/\sigma)$ and the incidence approaches $N(1-1/\sigma)d$.

Before the theoretical equilibrium level is reached, the magnitude or relative importance of the epidemiologic factors may change and thus define a new theoretical equilibrium level. Although the theoretical equilibrium levels may never be reached, the actual incidence will be close to the theoretical equilibrium since the approach to equilibrium is rapid in comparison to changes in the theoretical equilibrium (see the next section). Thus we can think of gonorrhea incidence as having a moving equilibrium where the movement is due to changes in epidemiologic factors.

Figure 1.2 in Chapter 1 shows that the reported incidence of gonorrhea increased each year from 1957 to 1975. Reported incidence in the United States increased by a factor of about 4 between 1960 and

1975. Because of the increased awareness of the seriousness of gonococcal infection in women and the screening program started in the
United States in 1972, the reporting of gonococcal infections in women
may be better in the seventies than in the sixties.   Changes in
reported cases in men may correspond better to changes in actual
incidence so that the actual incidence of gonorrhea from 1960 to 1975
may be increased by a factor of approximately 3 instead of 4 (Yorke,
Hethcote and Nold, 1978). The approach presented above would explain
this increase as the result of continuous changes in the epidemiologic
factors.   Factors often mentioned as possible causes of the increased
incidence include changes in sexual behavior and population mobility,
changes in gonococcal resistance to antibiotics and changes in methods
of contraception (WHO, 1978).

There is considerable evidence of changed sexual behavior in the
United States (and elsewhere).   Between 1967 and 1974, premarital
intercourse rates rose 300 percent for white women and 50 percent for
white men (NIAID, 1980).   A national survey of college students in
1976 showed rates of premarital coitus were 74 percent for both
sexes. Sexual activity among adolescents is clearly increasing. The
percentage of sexually experienced never-married women who have had
more than one sex partner increased from 38.5% in 1971 to 49.9% in
1976 (NIAID, 1980).

The demographic factors that correlate best with gonorrhea
incidence are age, race, marital status, socioeconomic status and
urban residence (WHO, 1978). Changes in demographic factors are often
ignored, yet they could cause significant changes in incidence.   For
example, if all other epidemiologic factors remained fixed, then a
change in the size of the high case rate age group should cause a
proportional change in gonorrhea incidence. Since the 18-24 age group
has the highest age specific case rates in the United States (American
Social Health Association, 1975; Zaidi et al, 1983) the increase in
size of this age group by a factor of 1.7 between 1960 and 1975
(Bureau of the Census, 1977) could have been one important cause of
the increase in reported gonorrhea incidence.   Reported incidence is
approximately ten times as high in the black population (WHO, 1978) on
a per capita basis and the black population in the 18-24 age group
increased by a factor of 1.9 between 1960 and 1975. Since the size of
the 18-24 age group is projected to decrease by a factor of 0.86 in
the United States from 1975 to 1990 (Bureau of the Census, 1977) a
corresponding decrease in gonorrhea incidence might be expected if
other epidemiologic factors remained constant.

Since 1975 the reported incidence of gonorrhea has been approximately constant. Although this suggests that an equilibrium level may have been reached, some epidemiologic factors have probably changed since 1975. Sexual activity among young people has probably continued to increase. The number of people in the high-risk age groups was expected to reach a maximum in 1983 (NIAID, 1980). The screening program initiated in 1972 has become larger and probably more effective. There has also been some indication that the fear of genital herpes has reduced the number of casual sexual contacts. Thus the almost constant observed incidence of gonorrhea is almost certainly a balance among factors which tend to increase and to decrease incidence. Of course, there is no way to measure the changes in these factors accurately enough to give quantitative predictions of the movement of the equilibrium.

2.3 Changes in Incidence: Rapid Response

Not only does the incidence of gonorrhea change as epidemiologic conditions change, but, in fact, the **incidence changes rapidly in response to epidemiologic changes.** As an example of rapid response, if all venereal disease clinics in a region were suddenly closed, then the actual incidence of gonorrhea in that region would increase sharply to a new level within a few months. However, if the epidemiologic change occurred in steps over a period of time, then the incidence changes would also occur over the period of time. As examples of rapid response, we cite the regularly observed increases in national reported cases approximately four weeks after Christmas and New Year's Day when most treatment facilities are closed and people have different patterns of interaction. These increases are of short duration so that the incidences quickly drop back to the usual levels. Another example is the seasonal changes in gonorrhea incidence due to seasonal changes in epidemiologic conditions (Yorke, Hethcote and Nold, 1978).

A crude estimate of the contact number can be obtained by using screening data from 1973-1975. As mentioned in section 1.3 it has been estimated that in 1975 the screening program for women discovered 10% of all gonococcal infections in women. If the 10% detected occur randomly among those with gonococcal infections, then the average infectious period is reduced by 10%. Using a trend line analysis on the reported incidence in men, it has been estimated that the effect of the screening program in 1975 was a 20% decrease in incidence in men (Yorke, Hethcote and Nold, 1978). If the incidence in women is also reduced by 20% and their duration is reduced by 10%, the prevalence is reduced by a factor of (0.8)(0.9) = 0.72 by the screening program. Using the simple model in section 2.1, we can equate two expressions for the prevalence with screening

$$1 - 1/(0.9\sigma) = 0.72(1-1/\sigma)$$

This equality yields a contact number σ of 1.40.

This value of σ yields estimates of the rates of response of $I(t)$. The linearization of [2.2] near $I=0$ is $dI/dt = (\sigma-1)I/d$, $I(0) = I_o$. Thus if the initial infective fraction I_o is very small, then $I(t) = I_o e^{(\sigma-1)t/d}$ so that the doubling time for $I(t)$ is $t_d = d(\ln2)/(\sigma-1)$. For example, if the contact number is $\sigma' = 1.4$ and the average duration d is 1 month, then the doubling time is 1.7 months. This estimates the doubling time if gonorrhea were introduced into a new population or if a new virulent strain such as PPNG were introduced. The concept of rapid response to epidemiologic changes is consistent with this rapid initial increase of the prevalence.

We can also obtain an estimate from σ of the speed of approach to equilibrium. If $I(t) = 1-1/\sigma + V(t)$, then the linearization of [2.2] around the equilibrium point $1-1/\sigma$ is $dV/dt = -(\sigma-1)V/d$. This differential equation has solution $V(t) = V_o e^{-(\sigma-1)t/d}$ so that the half life is $t_h = d(\ln2)/(\sigma-1)$. Thus the halving time near the endemic equilibrium point (the time a nonequilibrium prevalence takes to get half way to the equilibrium from its current position) equals the doubling time near the trivial equilibrium point. If $\sigma = 1.4$ and d = 1 month, then **the distance from the endemic equilibrium point is halved every 1.7 months.** Thus the rate of approach to the endemic equilibrium point is also rapid, which is consistent with the idea of rapid response.

CHAPTER 3

A REFINED MODEL FOR GONORRHEA DYNAMICS

The population which needs to be described by a model for the transmission of gonorrhea consists of those sexually active people who could be infected by their contacts. The model in Chapter 2 assumes that this population is homogeneous and uniform; however, that model is too simple since the population is really quite heterogeneous. A suitable model should allow for heterogeneity by incorporating many groups. The division into groups could be done according to differences in sex, sexual contact rates, sexual behavior, age, geographic location, socioeconomic status, etc. For example, some individuals are more active sexually than others in the sense that they have more frequent changes of sex partners. Some infected people, especially women, are essentially asymptomatic and do not seek treatment while others have symptoms which cause them to seek treatment.

In section 3.1 we develop a model for a population divided into n groups or subpopulations. We show that **either the disease dies out naturally for all possible initial levels or the disease remains endemic for all future time.** Moreover, the numbers of infectives and susceptibles in each group approach nonzero constant levels, which are independent of initial levels. The effects of changes in the parameter values (corresponding to epidemiological changes) on a disease can be determined by examining the resulting changes in the endemic equilibrium levels.

A method of determining the contact rates among groups by using a proportionate mixing assumption is described in section 3.2. With this assumption the threshold quantity which determines whether the disease dies out or remains endemic is an average contact number. Models with different groups are considered in subsequent chapters.

3.1 A Gonorrhea Model with n Groups

Assume that the population is divided into n groups and let N_i be the size of the subpopulation in group i. We assume that each group is homogeneous in the sense that all individuals in the group are similar. They should have the same rates of contact with new sexual partners, the same mean durations of infection and the same likelihood of acquiring infection during a sexual encounter with an infectious partner. We assume that individuals are either susceptible or infectious and that infectious individuals in a group have the same sexual behaviour and activity levels as susceptibles. Let $I_i(t)$ denote the prevalence in group i at time t so that the susceptible fraction in group i is $1-I_i(t)$. We measure time t in days.

Let λ_{ij} be the average number of adequate contacts (i.e., contacts sufficient for transmission) per unit time (one day) of an infective in group j with persons in group i. Since the susceptible fraction in group i is $1-I_i(t)$, the average number of susceptibles in group i infected per unit time by an infective in group j is $\lambda_{ij}(1-I_i(t))$ and the average number infected per unit time by N_jI_j infectives is $\lambda_{ij}N_jI_j(1-I_i(t))$.

Let d_i be the mean duration of infection in days for a person in group i. As in Chapter 2, we assume that each infective in group i has a fixed chance of recovering each day and that the probability is $1/d_i$. Thus the removal rate per day from the infectious class is N_iI_i/d_i. As noted in section 2.1, that this is equivalent to assuming that the durations of infection in group i have a negative exponential distribution (Hethcote and Tudor, 1980).

The differential equations for the model are

$$\frac{d}{dt}(N_iI_i) = (\sum_{j=1}^{n} \lambda_{ij}(N_jI_j)(1-I_i)) - N_iI_i/d_i \qquad [3.1]$$

with initial conditions $I_i(0) = I_{io}$ for i=1,2,...,n. The first term in each differential equation is the rate of new infections or incidence in group i and the second term is the removal rate due to recovery. Figure 3.1 shows the susceptible and infective compartments and the transfer rates between compartments.

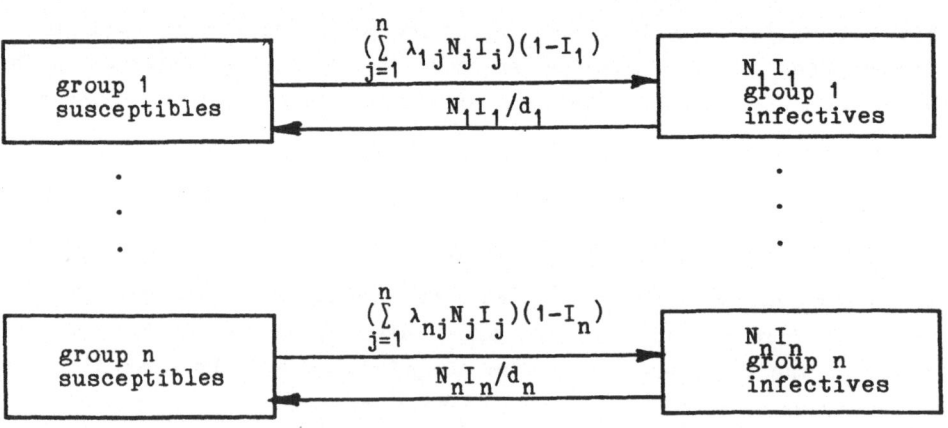

Figure 3.1 Flow diagram for the model [3.1]

Lajmanovich and Yorke (1976) proved that the model [3.1] is well posed. That is, unique solutions of [3.1] exist for all time, depend continuously on the initial data, and are always between 0 and 1. The nxn coefficient matrix A in the linearization of [3.1] is given by $A = L-D$ where $L = [\lambda_{ij}N_j]$ and D is a diagonal matrix with N_i/d_i as the entry in the ith row and column. Let $s(A)$ be the stability modulus of A, i.e., the maximum real part of the eigenvalues of A. They proved the following theorem.

THEOREM 3.1. Assume that the model is irreducible, that is, the population cannot be split into two subpopulations that do not contact each other. The solutions of [3.1] approach the equilibrium point at the origin if $s(A)<0$ and they approach a unique positive equilibrium point if $s(A)>0$, provided there is some infection in some group initially.

Thus **gonorrhea will die out if the parameter values are such that $s(A)<0$ and will approach an endemic steady state if** $s(A)>0$. One practical implication of the theorem above is that it allows us to focus on the positive equilibrium point and to see how it changes when parameter values change or when control procedures are added. Let $E_i>0$ be the equilibrium prevalence (the fraction of group i that is infectious at equilibrium). Thus the E_i are the solutions of the n simultaneous quadratic equations obtained when the right sides of [3.1] are set equal to zero. From the quadratic equations, the equilibrium incidence in group i is equal to the equilibrium prevalence E_i times the group size N_i divided by the mean duration d_i. Figure 3.2 shows the typical behavior of solution paths as they approach an endemic equilibrium point.

One of the striking features of Theorem 3.1 is the qualitative dynamical conclusion that equations [3.1] have a unique equilibrium point, either strictly positive or zero, which is the limit of every solution starting out from a state where infection is present. Hirsch (1984) has shown that this conclusion also holds for a generalization of equations [3.1]. In his differential equations, the incidence and removal terms are given by functions which satisfy certain conditions. His model is so general that it is not possible to give a procedure for deciding whether the equilibrium point corresponds to an endemic steady state or to die out of the disease. However, the generality of his model strongly suggests that any observed fluctuations in the incidence are not due to the intrinsic dynamics of the disease so that they must be due to fluctuations in

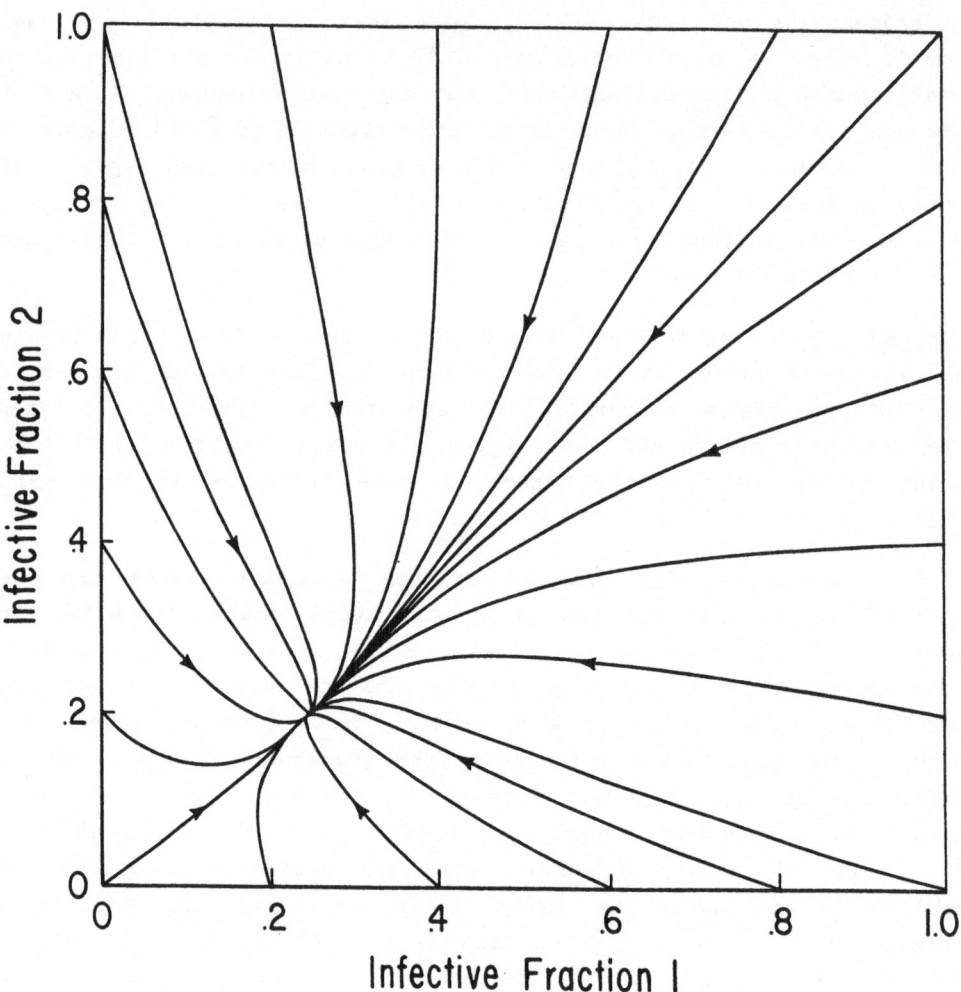

Figure 3.2.   Solution paths approaching the endemic equilibrium point
              when   s(A) > 0.

epidemiological or environmental factors or in reporting.

3.2 Proportionate Mixing Among Groups

The contact rates λ_{ij} in the contact matrix can be determined methodically by using some assumptions regarding the interactions of the groups. The "proportionate mixing" approach explained in Nold (1980) assumes that the number of adequate contacts between two groups is proportional to the relative sexual activities of the two groups. An encounter will refer to one or more episodes of sexual intercourse with a new partner. For example, if group 1 has 10% of all encounters and group 2 has 40% of all encounters, then in a proportionate mixing model, the fraction of all encounters which are between groups 1 and 2 is .10 × .40. The frequency of encounters is a better measure of sexual activity that is likely to transmit infection than the frequency of sexual intercourse, since encounters are new opportunities to become infected or to transmit the infection.

Let a_j be the activity level of group j, which is the average number of encounters of a person in group j per unit time. Thus $1/a_j$ is the average time between encounters for a person in group j. Let q_j be the probability that an infective in group j transmits the infection during an encounter with a susceptible, i.e., that there is an adequate contact. Let m_{ij} be the fraction of encounters made by an average infective of group j with persons in group i. Notice that the sum of each column in the mixing matrix M is 1. From these definitions it follows that the average number of adequate contacts per unit time of an infective in group j with different partners in group i is $\lambda_{ij} = a_i m_{ij} q_j$.

The average number of encounters per unit time is $A = \sum_{i=1}^{n} a_i N_i$. The fractional activity level of group i defined by $b_i = a_i N_i / A$ is a measure of the relative sexual activity of group i. Notice that $\sum_{i=1}^{n} b_i = 1$. The proportionate mixing assumption is that the **encounters of a person are distributed in proportion to the fractional activity levels**, i.e., $m_{ij} = b_i$.

The contact number k_j for group j, which is the number of adequate contacts made by a typical infective in group j during the duration of infection, satisfies $k_j = q_j a_j d_j$. If t_{ij} is the number of adequate contacts with group i of a group j infective during an average case, then $t_{ij} = \lambda_{ij} d_j = a_j m_{ij} q_j d_j = m_{ij} k_j$. The n × n matrix T = $[t_{ij}]$ is called the transmission matrix. In the proportionate mixing model, $t_{ij} = b_i k_j$.

The _average_ _contact_ _number_ for this model with proportionate mixing is $\bar{K} = \sum_{i=1}^{n} b_i k_i$, which is the weighted average of the contact numbers of the groups with the fractional activity levels used as weights. It is the average number of persons contacted by an average infective during the infectious period. We now prove that this average contact number is a threshold parameter which determines whether gonorrhea dies out ($\bar{K}<1$) or remains endemic ($\bar{K}>1$).

The characteristic equation for the transmission matrix T is $\det(T-\alpha I) = (-1)^n \alpha^{n-1}(\alpha-\bar{K}) = 0$. We assume below that T is irreducible, which again means that the whole population cannot be split into two subpopulations which do not interact with each other. The lemmas below are from Nold (1980).

LEMMA 3.2. If T is a square matrix with nonnegative elements, then T has a real, simple eigenvalue $p(T)$, called the Perron eigenvalue, which is equal to its spectral radius.

LEMMA 3.3. The outbreak eigenvalue $m_o = s(A)$ for [3.1] has the same sign as $r(T)-1$ where $r(T)$ is the spectral radius of T.

THEOREM 3.4. In the proportionate mixing model the solutions of [3.1] approach the origin if $\bar{K}<1$ and they approach a unique positive equilibrium if $\bar{K}>1$, provided there is some infection in some group initially.

PROOF. From the characteristic equation and Lemma 3.2, the Perron eigenvalue $p(T) = \bar{K}$ is equal to the spectral radius $r(T)$. By Lemma 3.3, $r(T) = \bar{K}<1$ is equivalent to the outbreak eigenvalue satisfying $m_o = s(A)<0$. The theorem now follows from Theorem 3.1.

We now develop some relationships that will be useful in later chapters. Using several definitions above, an algebraic manipulation leads to $\lambda_{ij}N_j/N_i = (k_i/q_i d_i)b_j q_j$ so that [3.1] becomes

$$\frac{dI_i}{dt} = (\sum_{j=1}^{n} b_j q_j I_j) \frac{k_i(1-I_i)}{q_i d_i} - \frac{I_i}{d_i} \qquad [3.2]$$

for $i = 1,2,\ldots,n$. This is a convenient form since the parameter values appearing are often available.

The endemic equilibrium prevalences E_i are found by setting the right sides of [3.2] equal to zero so they are the nontrivial solutions of

$$(\sum_{j=1}^{n} b_j q_j E_j)k_i (1-E_i)/q_i = E_i \qquad [3.3]$$

for i = 1,2,...,n. Define the <u>average</u> <u>equilibrium</u> <u>infectivity</u> h by

$$h = \sum_{j=1}^{n} b_j q_j E_j \ . \qquad [3.4]$$

The <u>fractional</u> <u>infectivity</u> of group j defined by

$$C_j = b_j q_j E_j / h \qquad [3.5]$$

measures the relative ability of group j to transmit the infection. From [3.3] and [3.4] we find that the endemic equilibrium prevalences E_i must satisfy

$$E_i = hk_i /(q_i + hk_i) \ . \qquad [3.6]$$

The equations [3.4] and [3.6] yield

$$\sum_{i=1}^{n} q_i b_i k_i /(q_i + hk_i) = 1 \qquad [3.7]$$

which is equivalent to an nth degree polynomial for h. For example if n=2, then the quadratic equation is

$$k_1 k_2 h^2 + [q_2 k_1 + q_1 k_2 - (q_1 b_1 + q_2 b_2)k_1 k_2]h - q_1 q_2 (\overline{R}-1) = 0 \ . \qquad [3.8]$$

The endemic prevalences are found from [3.6] using the positive root h of [3.8].

Since the incidence in group i is N_i times the summation terms in [3.2], the total incidence of the population per year divided by the population size (i.e., the number of cases per person per year) is

$$Y = \frac{365[\sum_{i=1}^{n} N_i (\sum_{j=1}^{n} b_j q_j E_j)k_i (1-E_i)/q_i d_i]}{\sum_{i=1}^{n} N_i} \ . \qquad [3.9]$$

Using [3.3] the number of cases per person per year satisfies

$$Y = \frac{365[\sum_{i=1}^{n} N_i E_i /d_i]}{\sum_{i=1}^{n} N_i} \ . \qquad [3.10]$$

CHAPTER 4

MODELING GONORRHEA IN A POPULATION WITH A CORE GROUP

In the early 1970s, the prevalent idea was that "gonorrhea is everybody's problem". It was recognized that everyone who was sexually active could get gonorrhea and, consequently, the screening program in the United States started in 1972 was designed to identify asymptomatic women by doing culture testing of as many women as possible. This screening program has been described in section 1.3.

In this chapter we study the core group: the group of individuals who are sexually very active and are efficient transmitters. The existence of a core group suggests that methods especially designed to identify and cure core members might be a more effective use of the resources available for gonorrhea control since the real objective of control is to prevent cases. A core-noncore model is developed in section 4.2 and parameter values such as sexual activity levels and average durations of infection are estimated. In the calculations the core is less than 2% of the population and consists of people who are 10 times as sexually active as noncore members. Remember that our measure of sexual activity is the frequency of new partners and that all individuals modeled are sexually active. Cases in the core are only 13% of the incidence; yet 16.7% of the encounters are with core members and 60% of all infections are directly caused by core members. Thus a small core group can be very important in the transmission of gonorrhea.

Various control procedures such as screening and contact-tracing are discussed in section 1.3. When screening and rescreening are compared in section 4.3, the calculations show that rescreening is approximately four times as effective per number of individuals tested as screening in reducing total incidence. As described in section 4.3 and in (WHO, 1978, p. 107), the National Strategy to Control Gonorrhea was revised in 1975 to include retesting and rescreening. The two strategies for contact investigation compared in section 4.4 are contact-tracing individuals named as infectors and contact-tracing infectees.

Recall from section 1.3 that if scientists eventually are able to develop a vaccine for gonorrhea, then it will probably give only short term immunity. Because of this limitation it is particularly valuable to examine ways of using a vaccine. The two vaccination strategies for potential vaccines compared in section 4.5 are vaccination of random individuals in the population at risk and vaccination of persons just after they have been treated for gonorrhea. Calculations in that section show that **post-treatment vaccination is about five times as effective per number of persons immunized as random vaccination** in the population.

The concept of a core is important in understanding gonorrhea dynamics from a clinical perspective. The quotation below is by R. K. St. John and J. W. Curran (1978) of the Venereal Disease Control Division of the Centers for Disease Control.

"Increasing emphasis is being placed on intervention strategies for the core population described by Drs. Yorke, Hethcote and Nold. Patients who have repeated infections in relatively short periods (three to six months) are clearly part of the core. Brooks, Darrow, and Day studied 7,347 patients from venereal disease clinics and retrospectively identified 492 patients who had had repeated infections. **This small number of patients was responsible for 21.6% of all cases of gonorrhea in the local county and 29.4% of all the cases seen in the clinic.** Membership in these high-risk groups constantly changes as variations in patients' sexual behavior lead individuals into or out of the group. Identification of these individuals while their risk of infection is high may have major impact on transmission of the disease. Studies are under way to determine which risk factors can be used to identify this group a priori so that attempts can be made through periodic screening to keep these patients free of disease for longer periods. [Emphasis added.]"

The quotation below is from a 1978 report of a World Health Organization scientific group (WHO, 1978, p. 116).

"The Group observed that, in the USA, the decision to carry out culture screening of non-symptomatic women was based on the assumption that the major cause of the gonorrhoea epidemic was the large reservoir of such women. This assumption is no longer in vogue in that country, and decisions for control are now based on the concept of core transmitters of disease, which postulates that a relatively small proportion of the population is contributing to the maintenance of the epidemic and that it is precisely this group of transmitters that is particularly important. Further disease models are needed for the development of innovative approaches to the problem of gonorrhoea control."

4.1 The Concept of a Core Group

If the contact number is greater than one and the initial suscep-tible fraction of the population is near one, then the initial infectee number is greater than one so that the prevalence for the disease will initially increase. The prevalence cannot increase indefinitely since it is bounded above by one. A factor which limits the prevalence of a disease is called a saturation factor. For diseases whose infection confers immunity, the saturation factor is acquired immunity. The average prevalence of such a disease is limited by the fact that some adequate contacts of an infective do not result in transmission since the contacted person is immune. Immunity acquired by infection or vaccination is the saturation factor for diseases such as measles, chickenpox, mumps, rubella, poliomyelitis, diphtheria and whooping cough. For influenza a mutation to a new strain can lead to a new epidemic since individuals may not be immune to the new strain.

Gonorrhea is an exception; most other directly-transmitted diseases confer significant levels of immunity. Since gonococcal infection does not appear to confer protective immunity or substantial resistance, acquired immunity cannot be a saturation factor for gonorrhea. An adequate contact of an infective will not result in transmission of gonococcal infection only if the contacted individual is also an infective. This is called the preemption effect since an already infectious individual cannot be infected by another infective. Strictly speaking the infected individual could acquire an additional strain. **Preemption seems to be the only possible saturation factor for gonorrhea** since infection does not confer immunity.

The effects of the screening program were used in our simplistic one population model in section 2.3 to estimate that the contact number σ is 1.40 so that the prevalence $1-1/\sigma$ before screening would be 0.29. If the initial prevalence is below 0.29, then the prevalence approaches this value monotonically, but does not exceed 0.29 because of the preemption effect. Thus saturation occurred in that model when the prevalence reached 0.29.

We now use crude estimates to calculate the prevalence in another way. We estimate that the actual yearly incidence of gonorrhea in the United States is 2.0 million and that the population at risk is approximately 20 million. If the average duration of infection is one month, then the number of cases at any given time is 166,667 which is less 1% of the at-risk population. Since less than 1% of the contacts of an average infective are also infectious, preemption is not an important limiting factor when the population is considered to be one large, uniform, homogeneously-infected population. If preemption at

1% were enough to stop incidence from increasing, then the large current screening program (which is estimated to discover and cure 10% of infectious women) would have caused gonorrhea to die out.

Of course, the population is not homogeneously infected and uniform since some individuals have more sex partners than others. Thus the preemption that limits gonorrhea must be occurring in a subset of the at-risk population. The sexually active population could be divided into subgroups according to sex, age, race, sexual practices, number of sex partners, etc. The population in the groups with high prevalence (with prevalences of at least 20%) are lumped together and called the core. **There is a significant preemption effect in the core.**

There is no sharp division of the population at risk into the core and noncore since the heterogeneous population is made up of many groups. However, it is convenient conceptually and computationally to think of the core and the noncore as the only groups in the sexually active population being considered. If half of the infected individuals (half of 166,667 people) were in the core and the prevalence for the core were 20%, then the core would have 416,667 members or about 2% of the 20 million people assumed to be at risk. Thus the core can be a small percentage of the population at risk.

The significance of the core can be determined by a thought experiment. We observe that prevalence in the noncore is small (~1%) so that no saturation is occurring there. In addition some cases there come from contacts with core members. We conclude that the contact number is less than one in the noncore. Suppose now that all individuals in the core were instantaneously cured and permanently immunized against gonococcal infection. Since cases infected by the core would no longer occur, the noncore prevalence would decrease and the infectee number for the entire population would decrease to a value less than one. Before the immunization the infectee number was greater than one for the core and less than one for the noncore; immediately after the immunization it is zero for the core and decreased for the noncore. Since there is no saturation in the noncore, the susceptible fraction of the population is approximately one so that the contact number is approximately equal to the infectee number. Thus the contact number for the noncore is less than one after the immunization so that gonorrhea would die out. In other words, since nonzero equilibrium prevalence required saturation in some group, if the core is immunized so that there is no saturation, then the equilibrium prevalence cannot be nonzero. Thus all cases are caused directly or indirectly by the core: **the core causes gonorrhea to remain endemic.**

Some efforts have been made to identify core groups by their characteristics (Rothenberg, 1982;Potterat, Rothenberg, Woodhouse et al, 1983). Homosexual men (Darrow et al., 1981; Judson et al., 1980) and prostitutes (Darrow and Pauli, 1983) may be core group members because of their frequent and anonymous sexual contacts, the social and legal impediments to their medical care, and their lack of referral of sex partners. An overall infection rate of 27% among street prostitutes was found in one city studied recently (Potterat et al., 1979). Gonorrhea was detected in 20% of women arrested for prostitution in Atlanta, Georgia (Conrad et al., 1981). There is some evidence that women with gonococcal Pelvic Inflamatory Disease are core group members because their sexual contacts are often infected and frequently asymptomatic (Wiesner and Thompson, 1980; Potterat et al., 1980). It is possible to identify high risk individuals by characteristics such as age, sex, race and census tract (Rothenberg, 1982). One study of possible core group members in a small community showed that 3% of the population was responsible for 27% of the gonococcal infections (Phillips, Potterat, Rothenberg et al., 1980). Intervention with high risk groups is an efficient use of resources since their importance in overall disease transmission is disproportionate to their numbers.

Groups with high rates of infection definitely are geographically clustered, often in an inner city (Potterat, et al. 1980; Wiesner, 1979; Rothenberg, 1982). The abstract of a paper, The Geography of Gonorrhea: Empirical Demonstration of Core Group Transmission, by Rothenberg (1983) is given below.

"The pattern of reported gonorrhea in Upstate New York (exclusive of New York City) in the years 1975-1980 is one of intense central urban concentration, with concentric circles of diminishing incidence. The relative risk for gonorrhea in these central core areas, compared to background state rates, is 19.8 for men and 15.9 for women, but as high as 40 in selected census tracts. Prevalence appears to approach 20% in some areas, the level postulated by current epidemiologic models for continuing endemic transmission. These core areas are characterized by high population density, low socioeconomic status and a male to female case ratio of one or lower. Contact investigation data suggest that sexual contact tends to exhibit geographic clustering as well. These observations provide support for narrow focusing of epidemiologic resources as a major disease control strategy."

4.2 The Core-Noncore Model

When part of the population differs from the rest in some epidemiologically significant way, it is desireable to consider the simplest model which concentrates on that difference. Here we use a model with a core group and a noncore group in order to determine the implications of having one group more active than the other. Two group models have the additional advantage that the equilibrium prevalences are easy to obtain with a hand calculator. However, the core-noncore model ignores distinctions between women and men and does not allow different groups with different durations of infection. A more refined model involving eight groups is considered in Chapter 6. There the equilibrium point is more difficult to calculate so a computer is used.

Here we divide the population into two groups based on the frequency of new sexual encounters. Let I_1 be the prevalence for the core (the very active group) and I_2 be the prevalence for the noncore (the active group). In order to do some calculations with the proportionate mixing model [3.2] with n=2, we must choose some parameter values. Let the average number of days between encounters be 5 days for core members and 50 days for noncore members so that the activity levels defined in section 3.2 are $a_1 = 1/5$ and $a_2 = 1/50$. Assume that the ratio N_1/N_2 of the sizes of the groups is 1/50 so that the core is about 2% of the sexually active population being considered. Let the average durations of infection, d_1 and d_2, both be 25 days. Here we assume that all new encounters are adequate contacts so that $q_1 = q_2 = 1$. Thus the core-noncore model is an initial value problem with the differential equations

$$\frac{dI_i}{dt} = \frac{k_i}{d_i} (b_1 I_1 + b_2 I_2)(1-I_i) - \frac{I_i}{d_i} \qquad [4.1]$$

for i = 1,2. A flow diagram for this model is given in figure 4.1.

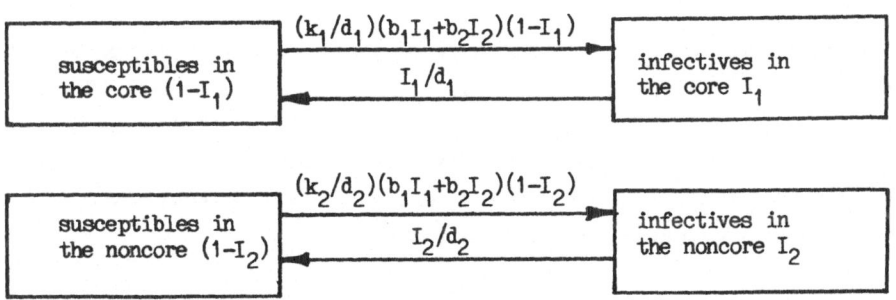

Figure 4.1 Flow diagram for the core-noncore model.

The difference between the groups is that core members are 10 times as active as noncore members. Without checking details, we might estimate roughly that a core member is 10 times as likely to become infected and has 10 times as many opportunities to transmit the infection while infected. Hence each core member might be expected to infect 100 times as many people as a noncore member. This calculation is not precise since it ignores the saturation in the core: an infective core member cannot be infected. We now see what the model predicts in detail.

Using definitions and equalities in section 3.2, we find that the contact numbers are k_1 = 5 and k_2 = 0.5 so that the average core member has adequate contacts with 5 people during the infectious period while each noncore person has adequate contacts with an average of 0.5 persons. The fractional activity levels are b_1 = 1/6 and b_2 = 5/6 so that 1 out of 6 encounters is with a core member and 5 out of 6 are with a noncore member. Thus 1/36 of the encounters are between two core members, 10/36 are between a core member and a noncore member and 25/36 are between two noncore members.

The average contact number $\bar{K}$ is 1.25 so that by Theorem 3.4, gonorrhea remains endemic and the prevalences approach the equilibrium values E_1 and E_2. From equation [3.8] the average equilibrium infectivity h is 0.078 so that the equilibrium prevalences are E_1 = 0.28 and E_2 = 0.038 from [3.6]. Thus 28% of the encounters of a susceptible with a core member result in infection while for encounters with a noncore member the figure is only 3.8%. From [3.5] the fractional infectivities are C_1 = 0.60 and C_2 = 0.40 so that for this model **60% of all infections are caused by core members.** Thus the core causes $1\frac{1}{2}$ times as many infections as the noncore even though he core is 50 times smaller. Thus for this model **a core member causes 75 times as many infections as a noncore member** (compare this with our original estimate of 100). From [3.10] the cases per person per year is Y = 0.62. Cases in the core group are 13% of the total incidence at equilibrium.

The calculations above show that **a small core group can be very important in the spread of gonorrhea.** The core is less than 2% of the population above and cases in the core are only 13% of the incidence; yet because core members are 10 times as sexually active as noncore members, 16.7% of the new encounters are with core members and 60% of all infections are caused by core members.

4.3 Screening and Rescreening Strategies

The following quotation is from the report of a World Health Organization scientific group (WHO, 1978, p. 105).

> "The mistaken assumption that every case of gonorrhoea is equally important for the spread of disease must be dispelled. Failure to treat gonorrhoea cases among groups with a high rate of disease transmission significantly limits the chances of success."

The general technique which we use is to change the differential equation to incorporate a control procedure such as screening. Then we solve for the new equilibrium. The drops in the equilibrium prevalence and incidence are a result of the control procedure. We can also calculate how many cases are prevented for each person discovered and treated via the screening program. The number of cases prevented includes all the cases in the chain of transmissions: primary, secondary, tertiary cases, etc.

First consider a strategy of screening individuals at random in the sexually active population. Let g be the fraction of the population being screened for gonorrhea by culture testing per day so that the yearly fraction screened is 365 times g. The fraction of those identified and cured by screening is gI_i for each group. When modified to include screening, the model [4.1] for the prevalences becomes

$$\frac{dI_i}{dt} = \frac{k_i}{d_i} (b_1 I_1 + b_2 I_2)(1 - I_i) - \frac{I_i}{d_i} - gI_i \ . \qquad [4.2]$$

for $i = 1,2$. The average contact number becomes

$$\overline{K} = \frac{b_1 k_1}{1 + gd_1} + \frac{b_2 k_2}{1 + gd_2} \qquad [4.3]$$

The equations for the equilibrium prevalences E_1 and E_2 are

$$\frac{k_i}{1 + gd_i} (b_1 E_1 + b_2 E_2)(1 - E_i) = E_i \qquad [4.4]$$

for $i = 1,2$. These equations are similar to [3.3] except that each k_i is replaced by $k_i/(1+gd_i)$.

We now calculate the effect of screening 1/2 of the sexually active population per year using the parameter values for the core-noncore model in section 4.2. Using $g = 1/730$ the quadratic equation yields h = 0.067 so that the equilibrium prevalences are E_1 = 0.24 and E_2 = 0.031. The cases per person per year calculated from [3.10] is Y = 0.53 which is a reduction of 13.9%. **For each 5.8 people screened there is a yearly reduction in incidence of 1 person.** The number of discoveries due to screening is $g(N_1E_1+N_2E_2)$ and the number screened is $g(N_1+N_2)$ so that the percent of those screened who are discoveries is 3.5%. The percent of people screened who are in the core is 13.2%.

Now consider a rescreening strategy where treated infectives return several weeks after their cure to be retested for gonococcal infection. These people who are rescreened could be infected by a new partner or reinfected by the old partner who is still infectious. Let f be the fraction of infectives who are rescreened and let the rescreening rates be proportional to the fractions removed from the infectious classes several weeks earlier. Since the several week delay is unimportant near the equilibrium, we assume that the fraction removed due to rescreening is the fraction f of recoveries screened times the recovery rate I_i/d_i times the prevalence I_i. The model [4.1] modified to include rescreening becomes

$$\frac{dI_i}{dt} = \frac{k_i}{d_i}(b_1I_1+b_2I_2)(1-I_i) - \frac{I_i}{d_i} - \frac{fI_i^2}{d_i} \qquad [4.5]$$

for i = 1,2. The equations for the equilibrium prevalences are

$$k_i(b_1E_1+b_2E_2)(1-E_i) = (1+fE_i)E_i \qquad [4.6]$$

for i = 1,2. These equations cannot be manipulated to give a quadratic equation like [3.8] for h, but they can be solved numerically for given parameter values.

Let us calculate the effect of rescreening 1/2 of the infected individuals using the parameter values given in section 4.2. Using f = 1/2, the equilibrium prevalences are E_1 = 0.206 and E_2 = 0.027 so that h = 0.057. The cases per person per year calculated from [3.10] is Y = 0.46 which is a reduction of 25.3%. Rescreening 1/2 of the infectives corresponds to rescreening a number equal to 23.1% of the total population per year. **For each 1.5 people rescreened, there is a yearly reduction in incidence of 1 person.** The number of discoveries due to rescreening is $f(E_1^2N_1/d_1+E_2^2N_2/d_2)$ and the number rescreened is $f(E_1N_1/d_1+E_2N_2/d_2)$ so that the percent of those rescreened who are discoveries is 5.1%. The percent of people rescreened who are in the core is 13.2%.

Thus **rescreening is approximately four times as effective per** number of individuals tested as screening in reducing incidence. The intuitive reason why rescreening is more effective than screening is that since rescreened individuals were infected before, they are more likely to be core group members and, consequently, more likely to be infectious again when rescreened. Moreover, discovering and curing a core member who can spread the infection to many others is more effective in reducing prevalence than discovering and curing a noncore member. Thus **rescreening is one method of focusing the culture testing on members of the core group,** who are the efficient transmitters.

In 1975 there was a change in the gonorrhea control program in the United States. It was recognized that there is a core group of efficient transmitters and that they are more likely to become reinfected shortly after treatment (Henderson, 1974b). The new strategies are described below (Henderson, 1974a).

"3. National Strategies to Control Gonorrhea

The major thrust of these strategies is the rescreening of gonorrhea patients after treatment for this disease. The elements of this overall strategy can be summarized into three points:

a. For infected persons both men and women:

   (1) Counsel to refer sex partners for examination and treatment; and

   (2) Counsel to return one week after treatment for a test-of-cure culture posttreatment culture) and 4-6 weeks after treatment for a rescreening culture.

b. For all patients with positive posttreatment followup cultures or with positive rescreening cultures, special efforts will be made to have their sexual partners referred for examination and treatment.

c. Improve clinical and laboratory services in both the public and private sectors to provide accurate diagnosis, effective therapy, and maximum utilization of services by persons at high risk of infection."

It is estimated that some of the 5% who are positive for gonorrhea on retesting after one week are treatment failures and some are reinfections (Henderson, 1975b). The first group is important because they may have PPNG and the second group is important because they and their sex partners may be more important transmitters, i.e., core

group members.   Rescreening after 4-6 weeks yields about 5-20% positive cultures in clinics.   Rescreening is designed to identify
individuals who are rapidly reinfected and, consequently, are in the
core group of more efficient transmitters.   The Director of the VD
Control Division of the Center for Disease Control stated that
(Henderson, 1974b) "the intended impact of these 'new' program
elements is to shift program resources from routine screening in
general populations to highly targeted testing and counselling in
populations with reinfections of recent origin."

The results of rescreening have varied from place to place.   In
those areas where it has been found to be effective, it has been
continued. The following quotation is illustrative (Miles, 1978).

> "Post-treatment culturing and rescreening at 4-6 weeks
> were first looked at as an activity that might be
> expendable to reduce laboratory support costs since it was
> done basically to monitor the effectiveness of therapy.
> However, it was discovered that a positivity rate of better
> than 5-6% was being achieved and that almost none of these
> patients were treatment failures.   This was a substantially
> higher rate than most providers obtained on initial
> culture, much less a reculture.   Most notable was the fact
> that almost no resources were being expended to identify
> these additional infections.   This group of patients was
> being reinfected between the time of treatment and
> reculture.   Without a doubt, these patients and their sex
> partners are among the most important transmitters of
> gonorrhea and by identifying them through rescreening and
> applying intensive epidemiology, we may more directly
> affect the incidence of gonorrhea than by all other control
> activities combined.   Therefore, our resources in Indiana
> were retargeted in 1976 to reach this important group of
> patients through rescreening in the venereal disease
> clinics."

4.4   Contact Investigation Strategies

Contact investigation or tracing procedures have been described
in section 1.3.   The quotation below is from Rothenberg (1982).

> "Resources available for contact interviewing and contact
> tracing are limited.   A coherent plan for targeting these
> resources, selecting priority patients--is economically
> mandated.   This approach is founded on a theoretical basis
> as well: high risk groups may be directly or indirectly
> responsible for most disease transmission (Yorke, Hethcote
> and Nold, 1978)."

Here we consider the effects of two different contact
investigation strategies.   These strategies are also analyzed using an
eight group model in Chapter 6.   One strategy is to contact-trace

individuals named as potential **infectees** or people to whom the disease may have been spread by the cases being considered. The other strategy is to contact-trace individuals named as **infectors**, i.e., the individuals from whom the cases being considered obtained their gonococcal infection. Anecdotal information suggests that infected individuals can usually correctly identify the person who infected them.

Assume that the daily number of people contact traced, found to be infectious and cured is a fraction f of the daily incidence. That is, for each current case, f other cases are cured as a result of contact tracing. If infectees are contact traced, then they are typical infectives so that the daily number contact traced are divided between the core and noncore in proportion to the incidences of the core and noncore. If infectors are traced, then those traced are divided between core and noncore in proportion to the numbers of infections caused by the core and noncore. The infectivity C_i defined in section 3.2 is the probability that the infection came from group i. The infectors are proven transmitters and so are more likely to be in the core than a random infected person.

In section 4.2 the calculations showed that the core accounts for 13% of the incidence so that 13% of randomly chosen cases would be core members. Thus 13% of the infectives found by contact tracing infectees would be core members. Since 60% of all infections were transmitted by core members, 60% of the infectives found by contact tracing infectors would be core members. It is clearly more important to identify and cure core members since they are the efficient transmitters. The typical noncore infective contacts 0.5 individuals when infectious while the typical core infective contacts 5, so there is a significant increase in cases prevented by curing a core infector, even if that individual is likely to be reinfected soon. The more detailed calculations below and in Chapter 6 verify this prediction.

If infectees are contact traced, then a fraction f of the incidences in each differential equation are removed by contact tracing. Thus the differential equations for the prevalences become

$$\frac{dI_i}{dt} = (1-f) \frac{k_i}{d_i} (b_1 I_1 + b_2 I_2)(1-I_i) - \frac{I_i}{d_i} \qquad [4.7]$$

for i = 1,2. The equilibrium prevalence equations are similar to [3.3] except that each k_i is replaced by $(1-f)k_i$. Using the parameter values in section 4.2, we model a control procedure in which 1% of the infectees (f = 0.01) are traced and cured. One reason for choosing f = .01 is so that it would be implementable in practice. Calculations show that h = 0.075 so that E_1 = 0.27 and E_2 = 0.036. The cases per person per year calculated from [3.10] is Y = 0.59 which is a reduction in incidence of 5.1%. **For each infectee contact traced and cured, there is a reduction in incidence of 5.34 people.**

If infectors are traced, then the number traced and cured is f times the total daily incidence. This number is divided between the core and noncore in proportion to the infectivities, $C_i = b_i I_i / (b_1 I_1 + b_2 I_2)$, which measure the relative ability of group i to transmit the infection (cf. equation [3.6]). Thus the fractional removal rates for i = 1,2 due to contact tracing are

$$\frac{C_i f(\text{incidence})}{N_i} = \frac{b_i I_i}{b_1 I_1 + b_2 I_2} \frac{f}{N_i} \left[\frac{N_1 k_1}{d_1} (1-I_1) + \frac{N_2 k_2}{d_2} (1-I_2) \right] (b_1 I_1 + b_2 I_2)$$

$$= a_i I_i f \frac{a_1 N_1 (1-I_1) + a_2 N_2 (1-I_2)}{A}$$

$$= \frac{k_i}{d_i} I_i f [1-(b_1 I_1 + b_2 I_2)]$$

Thus the differential equations for the prevalences become

$$\frac{dI_i}{dt} = \frac{k_i}{d_i} [b_1 I_1 + b_2 I_2][1 - (1-f)I_i] - (1+fk_i) \frac{I_i}{d_i} \qquad [4.8]$$

for i = 1,2. The equilibrium fractional prevalences satisfy

$$[k_i/(1+fk_i)][b_1(1-f)E_1 + b_2(1-f)E_2][1 - (1-f)E_i] = (1-f)E_i \qquad [4.9]$$

for i = 1,2. These equations are similar to [3.3] except that E_i is replaced by $(1-f)E_i$ and k_i is replaced by $k_i/(1+fk_i)$. Thus the quadratic equation approach given in section 3.2 can also be applied here. Using the parameter values in section 4.2, we find that if 1% of the infectors (f=.01) are traced and cured, then h = .069, $.99E_1$ = .247 and $.99E_2$ = .033. The cases per person per year is Y = 0.55 which is a reduction of 11.0%. **For each infector contact traced and cured, there is a reduction in incidence of 12.3 people.**

For tracing either infectees or infectors, f = .01 corresponds to successfully tracing and curing a number of infectives equal to 1% of the incidence. Thus according to this two group model curing a certain number of infectors by contact tracing is approximately twice as effective in reducing incidence as curing the same number of infectees by contact tracing. A model in Chapter 6 involving eight groups shows that tracing infectors is three times as effective as tracing infectees.

4.5 Vaccination Strategies

The efforts to develop a gonorrhea vaccine are described in section 1.3. The quotation below is from the report of the WHO scientific group (WHO, 1978).

> "One potential application of mathematical models for gonorrhoea is to predict the impact of preventive measures (e.g., vaccination) or of changes in case-finding or in efficacy of treatment (e.g., increased failure rates due to increased prevalence of β-lactamase-producing gonococci). Such projections could be useful in planning control programmes."

Here we investigate the effectiveness of two vaccination strategies. The general vaccination strategy is vaccination of individuals chosen at random from the population at-risk. The post-treatment strategy is vaccination of persons just after they have been treated for gonorrhea. Since it is likely that those who become immune due to vaccination will have only temporary immunity, let r be the average period of temporary immunity. Vaccine efficacy, which is the fraction of those vaccinated who become immune, can be much less than one.

For the general vaccination strategy let the daily rate of individuals in the ith group becoming immune due to vaccination be uN_iS_i. Let class R_i contain the people temporarily removed from the susceptible-infective interaction by immunity due to vaccination. The differential equation model [4.1] becomes

$$\frac{dI_i}{dt} = \frac{k_i}{d_i} (b_1I_1 + b_2I_2)S_i - \frac{I_i}{d_i}$$

$$\frac{dS_i}{dt} = - \frac{k_i}{d_i} (b_1I_1 + b_2I_2)S_i - uS_i + \frac{(1-S_i-I_i)}{r} + \frac{I_i}{d_i}$$

[4.10]

for $i = 1,2$. The fraction of the population which is temporarily immune is $R_i = 1 - S_i - I_i$. A flow diagram is given in Figure 4.2.

If we set the right sides of [4.10] equal to zero and add, then $S_i = (1-I_i)/(1+ru)$ so that the equilibrium points satisfy these equations. Thus the equilibrium prevalences satisfy

$$[\frac{k_i}{1+ru}] \; (b_1E_1 + b_2E_2)(1-E_i) = E_i \qquad\qquad [4.11]$$

or i = 1,2. These equations are similar to [3.3] except that each k_i is replaced by $k_i/(1+ru)$.

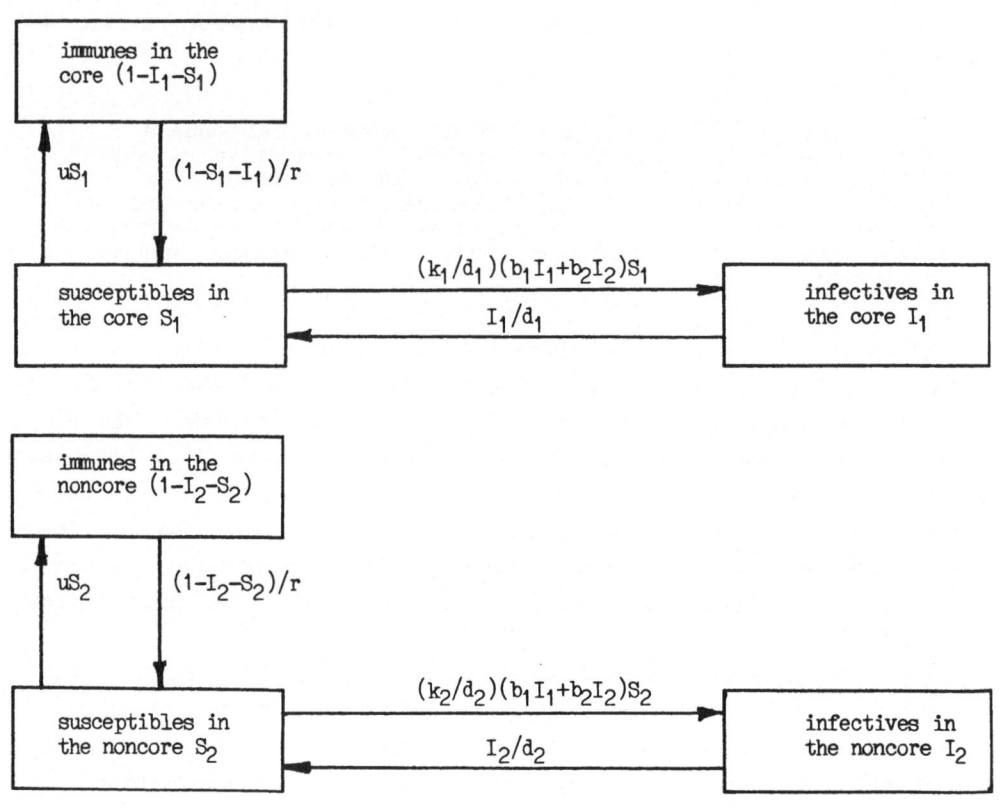

Figure 4.2 Flow diagram for the general vaccination strategy.

We now calculate the effect of randomly immunizing 1/20th of the population per year (u=1/7300) when the average period of temporary immunity is 6 months. Note that ur = 0.025 would also correspond to immunizing 1/40th of the population with average immunity of 1 year or to immunizing 1/10th with average immunity of 3 months. Using the parameter values for the core-noncore model in section 4.2, the equilibrium infectivity h is 0.070 so that the equilibrium prevalences are E_1 = 0.25 and E_2 = 0.033. The cases per person per year is Y = 0.54 which is a reduction of 12.4%. When 1/20th of the population is immunized by vaccination per year and the average immunity is 6 months, **there is a yearly reduction in incidence of 1.53 people for** each person immunized by general vaccination.

A vaccine could be very effective in controlling gonorrhea. Note that for a vaccine which gives an average immunity of 6 months, the calculations suggest that random immunization of 1/2 of the general population each year (ur > 0.25) would cause gonorrhea to disappear. Of course, vaccination would need to be continued forever at the same or a higher level to prevent an outbreak from an imported case.

We now analyze the post-treatment vaccination strategy in which the fraction f of the people treated are immunized by vaccination (possibly just after their treatment). We assume that the people treated are typical of the infectives. Let r be the average period of temporary immunity. The differential equations [4.1] become

$$\frac{dI_i}{dt} = \frac{k_i}{d_i}(b_1 I_1 + b_2 I_2)S_i - \frac{fI_i}{d_i}$$

$$\frac{dS_i}{dt} = - \frac{k_i}{d_i}(b_1 I_1 + b_2 I_2)S_i + \frac{(1-f)I_i}{d_i} + \frac{(1-S_i-I_i)}{r}$$

[4.12]

for i = 1,2. Again the fraction of the population which is temporarily immune is $R_i = 1 - S_i - I_i$. A flow diagram is given in Figure 4.3.

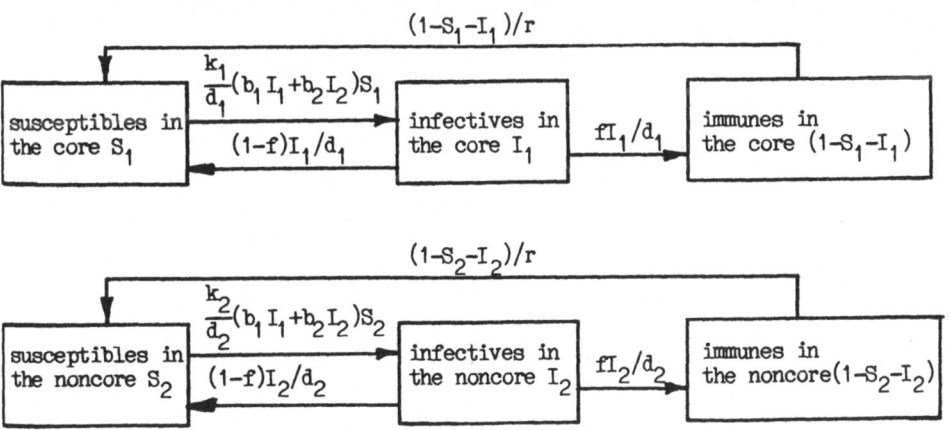

Figure 4.3 Flow diagram for the post-treatment vaccination strategy.

If we set the right sides of [4.12] equal to zero and add, then the equilibrium points must satisfy $S_i = 1-I_i(1+fr/d_i)$. Thus the equilibrium prevalences satisfy

$$k_i(b_1E_1 + b_2E_2)[1-E_i(\frac{1+fr}{d_i})] = E_i \qquad\qquad [4.13]$$

for $i = 1,2$. These equations can be made similar to the equilibrium equations [3.3] except that E_i is replaced by $(1+fr/d_i)E_i$.

Assume that $1/5$ of the infectives are immunized by vaccination and the average immunity r is 6 months. Note that $fr = 36.5$ days would also correspond to immunizing $1/10$ of the infectives with an average immunity of 1 year or to immunizing $2/5$ of the infectives with an average immunity of 3 months. Using the parameter values in section 4.2, the average equilibrium infectivity h is 0.032 so that the equilibrium prevalences are $E_1 = 0.11$ and $E_2 = 0.015$. From [3.10] the cases per person per year is $Y = 0.25$ which is a reduction of 59.4%. Immunizing $1/5$ of the infectives corresponds to immunizing a number equal to 5.0% of the total population per year. **For each person immunized by vaccination just after being treated** for a gonococcal infection, **there is a yearly reduction in incidence of 7.3 people.**

Hence **post-treatment vaccination is approximately 5 times as effective per person immunized as random vaccination.** Of course the random vaccination is restricted to the population being modeled where everyone is sexually active. Intuitively, post-treatment vaccination is better since core members (who are more often infected and are more efficient transmitters) are more likely to be temporarily immunized by post-treatment vaccination.

CHAPTER 5

MODELING GONORRHEA TRANSMISSION IN A HETEROSEXUAL POPULATION

Although sexually transmitted diseases are a major health problem among homosexuals, there is little transmission between the homosexual population and the heterosexual population (WHO, 1978; Wiesner and Thompson, 1980). Here we concentrate on a heterosexual population subdivided into women and men since the characteristics of gonorrhea are different for the two sexes. Although the groups of women and men considered here are highly active sexually, the results are more general since it was shown in Chapter 4 that changes in prevalence in the noncore group are directly related to changes in prevalence in the core group.

From the quarterly reported incidence of gonorrhea in women and men shown in figure 5.1, it is seen that gonorrhea incidence has a small but distinct seasonal oscillation. The quarterly incidence smoothed by using seasonal indices derived from the data is shown in figure 5.2. Notice that the seasonal oscillation is less than 10 percent. In this chapter we investigate the implications of the epidemiological differences between women and men and analyze the nature of the seasonality.

In the female-male model derived in section 5.1, a contact number determines whether the disease dies out or remains endemic. In section 5.2 many sources are used to estimate the parameter values and then the sensitivity of the prevalences and incidences at equilibrium to changes in parameter values is investigated. It is shown that the prevalences depend primarily on the contact number while the yearly incidences depend on the contact number and the average durations of infection. When screening programs are compared in section 5.3, it is found that because women have a longer average duration of infection, screening women is much more efficient than screening men. The effectiveness of screening women is proportional to the average duration of infection for women.

Epidemiologists have not understood why the peak incidence of gonorrhea occurs each year in August to October. In section 5.4 a model with small oscillations in the contact rates is analyzed mathematically using a perturbation analysis. The observed 6% seasonal oscillations in incidence in women and 10% oscillations in incidence in men may be due to reasonably small (5% to 7%) oscillations in the contact rate. From the analysis of the model, it appears that the

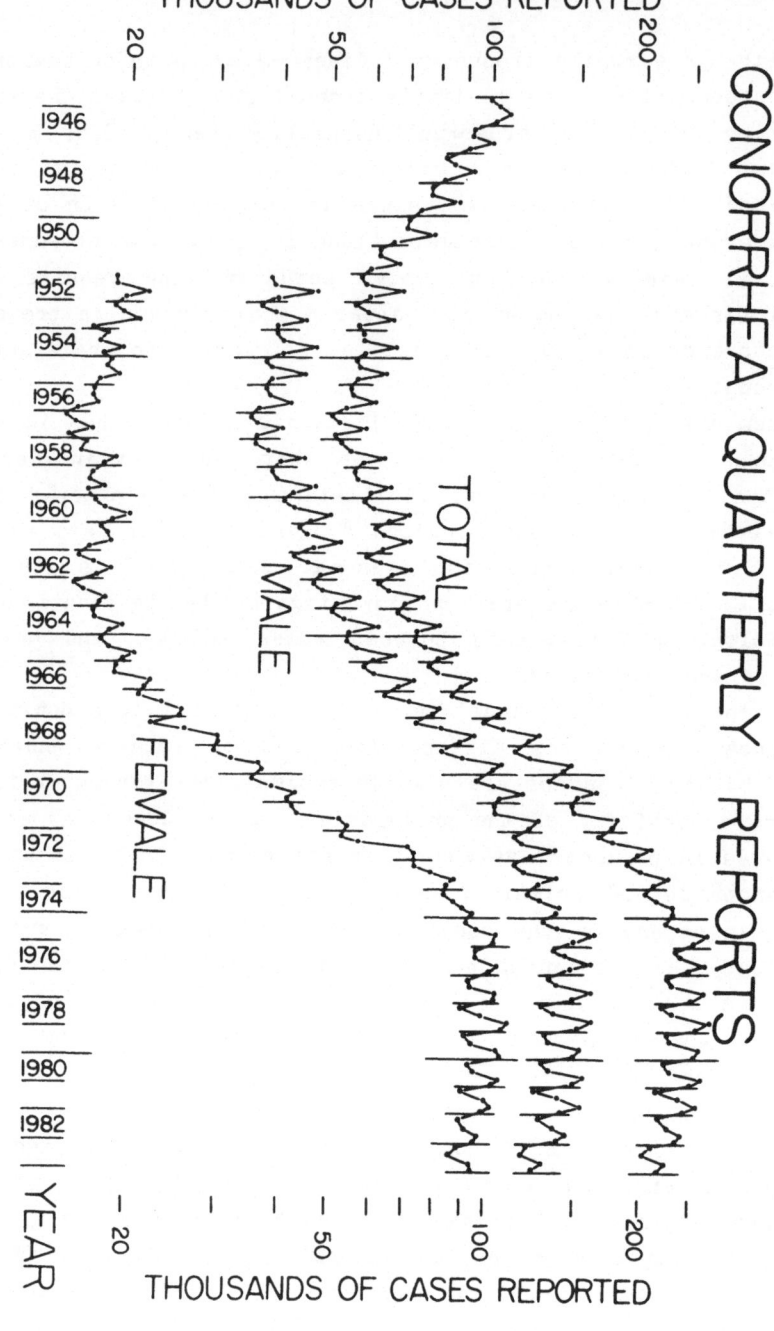

Figure 5.1. Reported cases of gonorrhea in women and men in the United States.

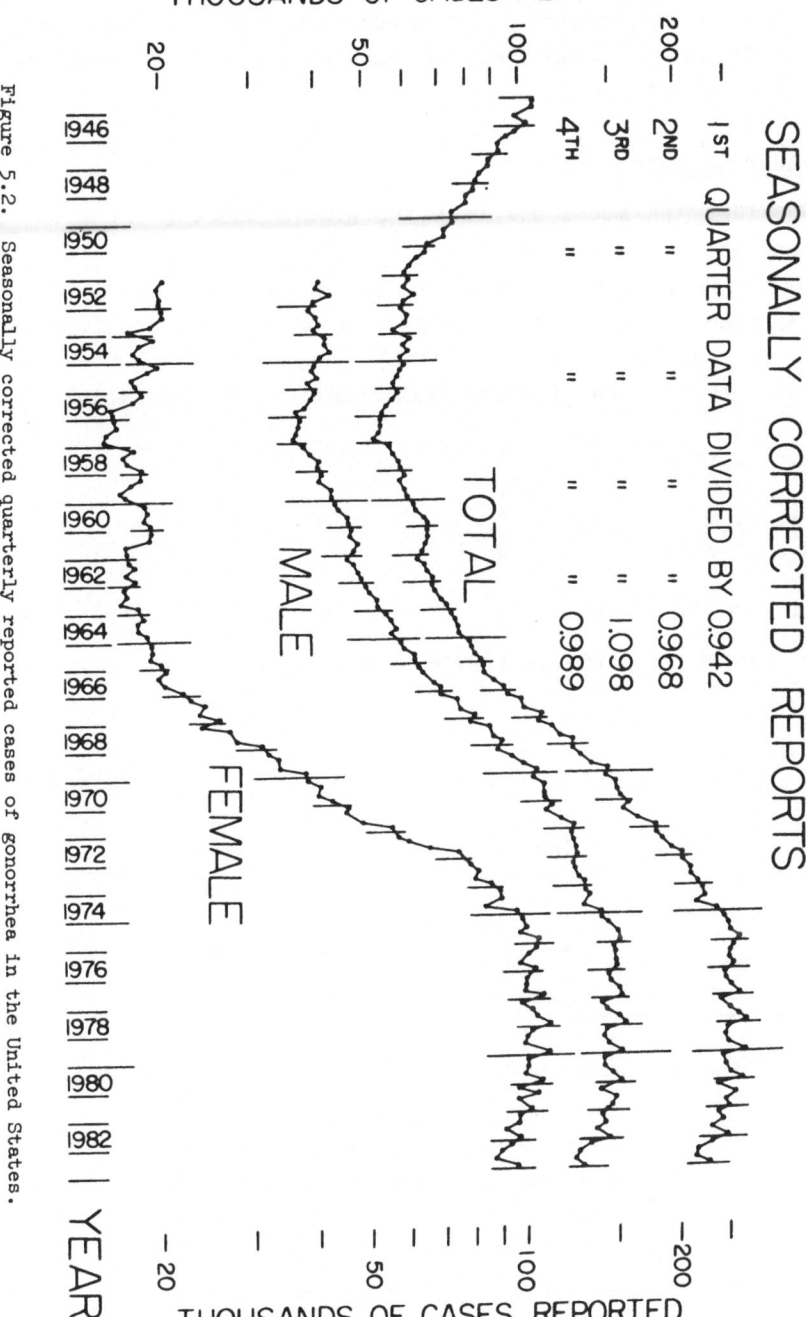

Figure 5.2. Seasonally corrected quarterly reported cases of gonorrhea in the United States.

observed peaks in August to October may be due to a peak contact rate
about two months earlier. This prediction that the peak contact rate
is probably in the summer months agrees with the data and the intui-
tion of epidemiologists.

5.1 The Female-Male Model

Consider the model 3.1 with two groups where group 1 consists of
women and group 2 consists of men. Since it is assumed that there is
only heterosexual transmission of gonococcal infection, the contact
rates λ_{11} and λ_{22} are zero. Thus this is <u>not</u> a proportionate mixing
model. Indeed, the model is formally the same as a host-vector model
(Hethcote, 1976). The differential equations for the model are:

$$\frac{dI_1}{dt} = (\frac{\lambda_{12}}{r})(1-I_1)I_2 - \frac{I_1}{d_1}$$
$$\frac{dI_2}{dt} = (\lambda_{21}r)(1-I_2)I_1 - \frac{I_2}{d_2} \qquad [5.1]$$

where $r = N_1/N_2$ is the ratio of the female to male population sizes.
A flow diagram is given in figure 5.3.

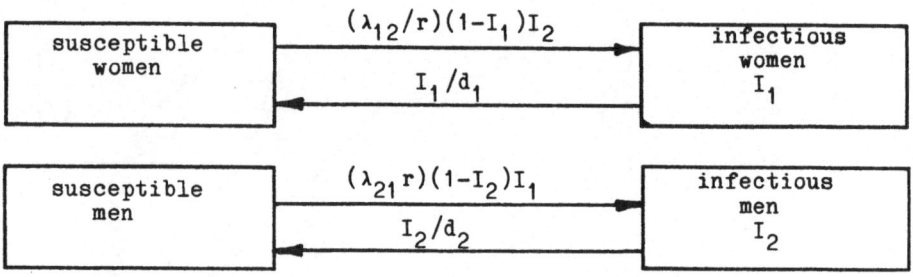

Figure 5.3 Flow diagram for the female-male model.

Let a_1 and a_2 be the activity levels of women and men, which are
the average daily rates of new encounters by women and men,
respectively. Let q_1 be the probability that there is an adequate
contact by an infectious woman during a new encounter and q_2 be the
analogous probability for an infectious man. Using these definitions,
the contact rates are $\lambda_{12} = q_2 a_2$ and $\lambda_{21} = q_1 a_1$. The contact number
k_1 for women, which is the average number of adequate contacts by an
infectious woman during her infectious period, satisfies
$k_1 = \lambda_{21} d_1 = q_1 a_1 d_1$. The contact number k_2 for men satisfies

$k_2 = \lambda_{12}d_2 = q_2 a_2 d_2$.

The second generation contact number $K = k_1 k_2$ in the theorem below determines whether the disease fades out. Since K is the product of the female contact rate and the male contact rate, it has the following interpretation: K is the average number of women (second generation) adequately contacted by men (first generation) who were adequately contacted by an average infectious woman during her infectious period. It also has a symmetric interpretation by switching the roles of women and men. See Hethcote (1974) or Lajmanovich and Yorke (1976) for a proof of the following threshold theorem.

THEOREM 5.1   If $K<1$, then the solutions $I_1(t)$ and $I_2(t)$ of [5.1] approach 0 as t approaches ∞ (i.e., fade out case). If $K>1$, then for positive values of $I_1(0)$ or $I_2(0)$, the solutions $I_1(t)$ and $I_2(t)$ of [5.1] approach E_1 and E_2, respectively as t approaches ∞ (i.e., endemic case) where the female and male prevalences at the endemic equilibrium are

$$E_1 = \frac{K-1}{K+(\lambda_{21}r)d_2} , \qquad E_2 = \frac{K-1}{K+(\lambda_{12}/r)d_1} . \qquad [5.2]$$

This theorem has an intuitive interpretation. If the average infectious woman infects less than one other second generation woman even at low prevalence levels, then gonorrhea dies out. If she infects more than one, then gonorrhea remains endemic and the prevalences approach equilibrium levels. It can be verified algebraically that at the endemic equilibrium, the infectee number $KS_1 S_2 = k_1 k_2 (1-E_1)(1-E_2)$ is 1 as predicted in section 1.5. The female cases per woman per year at equilibrium Y_1 is equal to the prevalence E_1 times the population size N_1 divided by the duration d_1 given in years. The definition of Y_2 is analogous.

Since the number of encounters of women must equal the number of encounters by men, $a_1 N_1 = a_2 N_2$. This relationship can be used to reduce the number of parameters appearing in the four coefficients in [5.1] from five (λ_{12}, λ_{21}, r, d_1, d_2) down to four. Define the contact effectiveness ratio e to be q_1/q_2. Then

$$\frac{\lambda_{21}r}{\lambda_{12}/r} = \frac{q_1 a_1 N_1 r}{q_2 a_2 N_2} = er$$

so that

$$K = (\lambda_{12}/r)(\lambda_{21}r)d_1d_2 = (\lambda_{12}/r)^2(er)d_1d_2 = [(\lambda_{21}r)^2/er]d_1d_2$$

and

$$\lambda_{21}r = \left[\frac{K(er)}{d_1d_2}\right]^{1/2} \qquad \lambda_{12}/r = \left[\frac{K}{(er)d_1d_2}\right]^{1/2} \qquad [5.3]$$

Thus the four coefficients in [5.1] now depend on the four convenient parameters d_1, d_2, K and er. Since K, e, and r have a more direct epidemiologic interpretation than λ_{12} and λ_{21}, better estimates of them can be made from available data.

5.2 Parameter Estimation and Sensitivity Analysis

Since it is not possible to estimate the ratio of the population sizes N_1 and N_2 of the women and men at risk, we simply assume that r = N_1/N_2 is 1. As described in section 1.2 the probability of transmission of gonococcal infection during one sexual intercourse by an infectious woman is about 0.2 to 0.3 while the corresponding probability of transmission by an infectious man is 0.5 to 0.7 (Wiesner and Thompson, 1980; Rein, 1977). Thus the probability of transmission in n sexual intercourses increases as n increases and can be estimated to be $1-(0.75)^n$ for an infectious woman and $1-(0.4)^n$ for an infectious man though in fact the n events are not truly independent. If encounters consisted of exactly one, two or three sexual intercourses, then the contact effectiveness ratio e = q_1/q_2 would be approximately 0.42, 0.52, 0.62, respectively. Since some encounters involve only one sexual intercourse and some involve several, the value used for er is 0.5.

The average durations of infection can be calculated as a weighted average of the average durations of symptomatics and asymptomatics. Estimates of periods of infection are 3-45 days for symptomatic women, 3-12 months for asymptomatic women, 3-30 days for symptomatic men and 3-6 months for asymptomatic men. Moreover, approximately 60% of cases in women are asymptomatic and 10% of cases in men are asymptomatic (Wiesner and Thompson, 1980; Kramer and Reynolds, 1981). Realistically, there is no way to obtain highly reliable estimates of these values. Using average durations of infection of 8 days for symptomatic women and 128 days for asymptomatic women, the weighted average duration for women is 80 days. Using an average duration of 8 days for symptomatic men and 128 for asymptomatic men, the weighted average duration for men is 20 days.

The contact number K is greater than 1 since gonorrhea is endemic. It cannot be close to 1 since small changes in sexual behavior or in health care delivery would then cause large changes in incidence and large changes have not been observed (Yorke, Hethcote and Nold, 1978). Here the contact number K is taken to be 1.4 as in section 2.3.

The ratios of reported female cases to reported male cases in the United States for the calender years from 1964 through 1980 were: .33, .33, .32, .33, .33, .35, .37, .42, .52, .65, .68, .68, .68 , .68, .70, .70, and .69 (Blount, 1979). The increase in this ratio in the early seventies is obvious and coincides with the increase in nationwide screening and with the awareness of the importance of finding infected woman. Epidemiologists believe that the increase is due to the increased searching out and identification of infective women by the screening program. Studies involving contact investigation have been used to estimate the ratio of actual female to actual male cases, but the estimates are not consistent so that the ratio of actual incidences is unknown (Rein, 1977). Consequently, for a model to be satisfactory, we require the ratio (using equilibrium values) to lie between 0.6 and 1.0. The current ratio of reported cases is 0.69. In fact this ratio may vary from population to population. Our best estimate of the parameters of the model is parameter set number 1 in Table 5.1. The uncertainty in this "baseline" parameter set requires the examination of the other sets in that table.

The sensitivity of the prevalences and yearly incidences in the model in section 5.1 to changes in parameter values is now investigated. Table 5.1 shows the prevalences and incidences for the baseline parameter set (number 1) and for modified parameter sets. At the endemic equilibrium for the baseline parameter set 1, 22% of the women and 8.4% of the men have gonorrhea at a given time so that the susceptible fractions are .78 for women and .916 for men. For the baseline parameter set 1 the contact numbers are k_1 = 1.67 for women and k_2 = 0.84 for men. The average number of transmissions at the endemic equilibrium by an infectious woman is 1.53 and by an infectious man is .65 so that the infectee number is 1. For the baseline parameter set 1 the prevalence in women is above 0.20 but the prevalence in men is not so that according to the criterion in section 4.1, the women form the core group in this model. For the baseline parameter set 1 the yearly incidence in women is 0.65 times the yearly incidence in men which is consistent with the ratio 0.69 of reported incidences.

From parameter sets 1-7 we see that the prevalences E_1 and E_2 depend primarily on the contact number K and only slightly on d_1, d_2 and er. On the other hand, the yearly incidences are strongly dependent on the durations. Doubling both durations d_1 and d_2 as in parameter set 2 does not change the prevalences, but it does halve the yearly incidences. At first glance it may seem strange that the prevalences are not changed. This is because the contact number remains unchanged. In effect, increasing the duration automatically decreases the number of contacts per day for both men and women. As seen in parameter set 7, the value of the parameter er influences the distribution of the prevalence between women and men. Some of the qualitative observations above can be deduced from equation [5.2] and [5.3] and hold for all parameter values.

Although the estimates in this section of the parameters are subject to uncertainty and the model in section 5.1 involves simplifications, the model and baseline parameter set 1 are accurate enough to obtain comparisons and estimates in subsequent sections.

TABLE 5.1

Equilibrium Prevalences and Yearly Incidences
for Various Parameter Sets

Parameter set	1	2	3	4	5	6	7
Duration d_1	80	160	160	80	80	80	80
Duration d_2	20	40	20	40	20	20	20
Contact number K	1.4	1.4	1.4	1.4	2.	1.2	1.4
Parameter er	.5	.5	.5	.5	.5	.5	1.
λ_{21} r	.021	.010	.015	.015	.025	.019	.030
λ_{12}/r	.042	.021	.030	.030	.050	.039	.030
Equilibrium							
Prevalence E_1	.220	.220	.236	.201	.400	.126	.201
Prevalence E_2	.084	.084	.065	.106	.167	.047	.106
Y_1=yearly cases per woman	1.004	.502	.538	.916	1.825	.575	.916
Y_2=yearly cases per man	1.538	.769	1.190	.969	3.042	.849	1.938

5.3 Screening Women and Men

We consider the effect of screening as a gonorrhea control procedure by modifying the female-male model in section 5.1 to include screening. Let C_1 and C_2 be the fraction of the women and men that are screened for gonorrhea per day. Assume that the screened fraction is a random sample so that its mixture of susceptibles and infectives is typical of the populations being considered. Thus we assume that the fractions $C_1 I_1$ and $C_2 I_2$ are treated and removed per day from the respective infective classes.

When the differential equation model [5.1] is modified to include screening, it becomes

$$\frac{dI_1}{dt} = (\frac{\lambda_{12}}{r})(1-I_1)I_2 - \frac{I_1}{d_1} - C_1 I_1$$

$$[5.4]$$

$$\frac{dI_2}{dt} = (\lambda_{21} r)(1-I_2)I_1 - \frac{I_2}{d_2} - C_2 I_2$$

A flow diagram is given in figure 5.4.

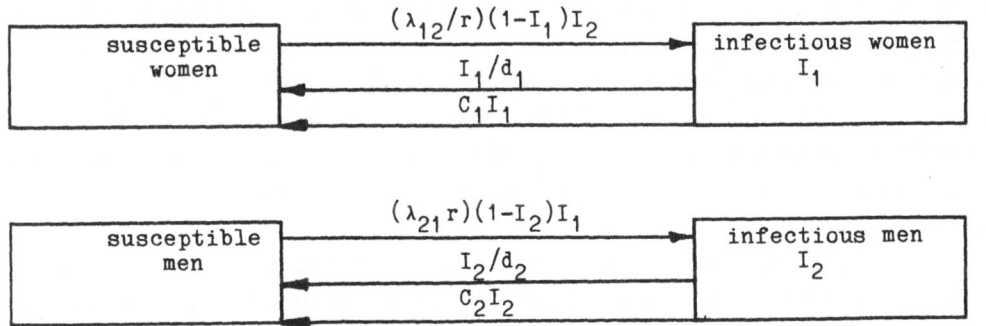

Figure 5.4. Flow diagram for the female-male model with screening.

Thus the net effect of screening is to decrease the average duration of infection and, consequently, the contact number. If d_1^s and d_2^s denote the average female and male durations in the presence of screening, then

$$d_i^s = [1/d_i + C_i]^{-1}.$$

Screening women is much more effective in reducing average

duration than screening men at the same rate because the average duration of infection is longer for women. For example, assume that 25% of the women and men are screened per year so that $C_1 = C_2 = .25/365$. If $d_1 = 80$ days and $d_2 = 20$ days, then $d_1^s = 75.8$ which is a 5.2% decrease and $d_2^s = 19.73$ which is a 1.4% decrease. In this case the percentage decrease in average duration is about 4 times greater for women than for men. The average duration of women is larger than for men since more women are asymptomatic. Indeed, the national screening program in the United States screens only women in an attempt to identify asymptomatic women. In the remainder of this section we will only consider screening of women.

Table 5.2 shows some calculated values of prevalences at equilibrium and yearly incidences for various yearly screened fractions 365 C_1 of women. Parameter sets 1 and 3 from Table 5.1 were used as baseline parameter sets in computing the percentage changes. For parameter set 2 in Table 5.2 interception and cure of 5.5% of the female infectives in each 80 day period (the average duration in women) shortens the average duration in women by 5.2% so that the prevalence in women is reduced by 14.8% and the prevalence in men is reduced by 13.7%. Since both the prevalence and durations are decreased in women, the female cases per woman per year is reduced by only 10.2%. Since the average duration in men is unchanged, the male case per man per year is reduced by 13.7%. In parameter set 5, screening women an average of 2 times per year causes gonorrhea to die out. It is clear in Table 5.2 that screening a given fraction of women is more effective in reducing prevalences and incidences when the original average duration for women is 160 days than when it is 80 days.

As described in section 5.2 the ratio of female to male reported cases increased from .33 to about .69 when the screening program was started. We now investigate possible causes of this increase.

It is estimated that 7 out of the 8 million culture tests each year are screening tests and the balance are diagnostic tests (Yorke, Hethcote, and Nold, 1978). The number of women in the United States between ages 19 and 29 is about 28 million so that the female population at risk is probably less than 28 million. It could be something like 7 million or 14 million. Thus the fraction of the female population at risk screened each year might be about 25% or 50% or 100%. In the calculations for parameter sets 1-4 in Table 5.2, the ratio of Y_1 to Y_2 increases from 0.65 with no screening to 0.68 with 25% screening to 0.71 with 50% screening to 0.76 with 100%

screening. These calculations suggest that a screening program for women could cause a slight increase in the ratio of incidences in women to men, but could not cause this ratio to double. Hence, the explanation of the observed doubling of this ratio is probably that given by most epidemiologists; namely, that the doubling of the ratio of incidences in women to men is due to increased case finding in women.

TABLE 5.2

Equilibrium Prevalences and Yearly Incidences for Various
Yearly Screened Fractions $365C_1$ of Women.

Parameter set	1	2	3	4	5	6	7	8
duration d_1	80	80	80	80	80	**160**	**160**	**160**
duration d_2	20	20	20	20	20	20	20	20
contact number K	1.4	1.4	1.4	1.4	1.4	1.4	1.4	1.4
parameter er	.5	.5	.5	.5	.5	.5	.5	.5
$365C_1$	0	.25	.5	1	2	0	.25	.5
prevalence E_1	.220	.187	.156	.095	0	.236	.168	.103
prevalence E_2	.084	.073	.061	.038	0	.065	.047	.029
Y_1=yearly cases per woman	1.004	.902	.780	.527	0	.538	.425	.286
Y_2=yearly cases per man	1.538	1.327	1.117	.695	0	1.190	.864	.538
% change in Y_1	0	-10.2%	-21.4%	-47.5%	-100%	0	-21.0%	-46.8%
% change in Y_2	0	-13.7%	-27.4%	-54.8%	-100%	0	-27.4%	-54.8%

5.4 Seasonal Oscillations in Gonorrhea Incidence

The reported incidence of gonorrhea in the United States has oscillated seasonally ever since data collection was started in 1919 (Cornelius, 1971; Jones, 1978). Seasonality of reported incidence has also been observed in Austria, Sweden and Bulgaria (Rein, 1977). The maximum incidence in the United States, which has always occurred in August to October, is at least 20% higher than the minimum incidence, which has always occurred in February to May (Wiesner and Thompson, 1980). Cornelius (1971) used quarterly data from 1950 through 1968 to calculate seasonal indices for gonorrhea incidence. The seasonality of reported incidence using quarterly data from 1946 to 1977 is shown in figure 5.1. The smoothness of the quarterly incidence data corrected by seasonal indices in figure 5.2 shows the regularity of the seasonal variation of gonorrhea incidence. The median of the weekly reported cases for 1966-1980 are shown in figure 5.5. Even though the median is very erratic the low point seems to be in the winter or spring and the peak is between August and October.

The reason for the seasonality of reported gonorrhea is unknown. It does not seem to be due to variations in reporting (Cornelius, 1971; Rein, 1977). Legitimate and illegitimate conceptions show the opposite seasonal pattern as gonorrhea. Although syphilis incidence does not seem to vary seasonally, incidence of nonspecific urethritis has the same seasonal pattern as gonorrhea (Rein, 1977). Similar seasonal case patterns have been observed for both sexes, for public and private cases, for large and small cities, for rural areas, and for cities with temperate and severe winters (Cornelius, 1971). The seasonality could reflect seasonality in susceptibility to gonorrhea or in the virulence of the gonococcus or increased use of antibiotics in winter months (Wiesner and Thompson, 1980). W. W. Darrow at the Centers for Disease Control predicted that the peak contact rate for gonorrhea should occur in the summer when students and other people often move and change sex partners. Why the peak incidence of gonorrhea occurs in August to October has baffled epidemiologists. This late peak is explainable by this model.

A perturbation analysis is now used to determine an approximate solution for a small oscillation in the contact rates. Some readers may wish to skip the detailed mathematical analysis and go directly to the conclusions at the end of the section. Note that Aronsson and Mellander (1980) showed that if the general model [3.1] is modified so that the contact rates and removal rates are periodic, then above the threshold there is a unique nontrivial periodic solution, which is globally asymptotically stable. Our analysis below yields estimates and further information regarding the unique periodic solution of the female-male model.

Seasonality is introduced into the model [5.1] by assuming that the contact rates vary seasonally so that λ_{12} and λ_{21} are both multipled by $1 + \epsilon \sin wt$ where ϵ is the relative amplitude of the perturbation and the frequency w corresponds to a period of one year. The model [5.1] becomes

$$\frac{dI_1}{dt} = (\frac{\lambda_{12}}{r})(1 + \epsilon \sin wt)(1-I_1)I_2 - \frac{I_1}{d_1}$$

$$\frac{dI_2}{dt} = (\lambda_{21}r)(1 + \epsilon \sin wt)(1-I_2)I_1 - \frac{I_2}{d_2}$$

[5.5]

MEDIAN OF WEEKLY REPORTED CASES OF GONORRHEA FROM 1976 TO 1980 IN THE UNITED STATES

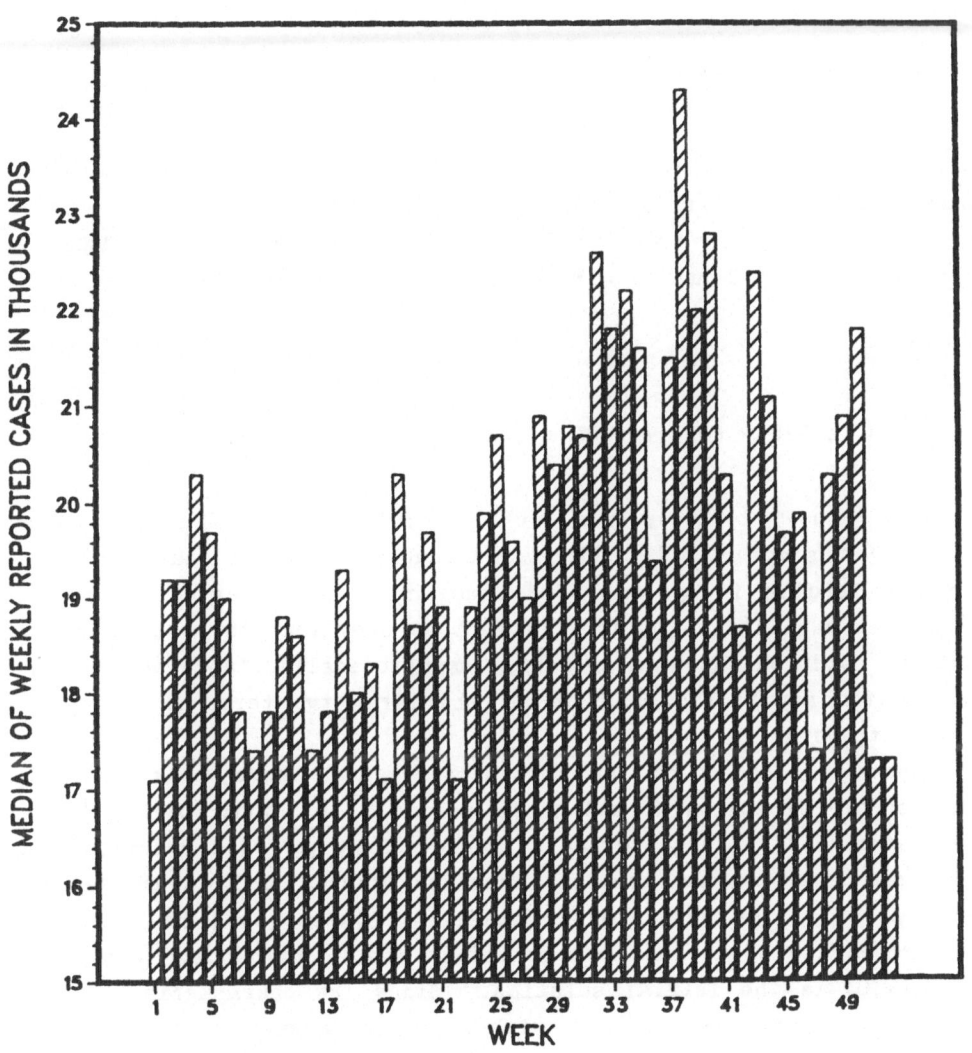

Figure 5.5.   Median of weekly reported cases of gonorrhea from 1976 to 1980 in the United States.

The dimensionless and scaled form of this model is

$$\frac{dI_1}{d\tau} = \mu(1 + \epsilon \sin \psi\tau)(1-I_1)I_2 - I_1$$

$$\frac{dI_2}{d\tau} = \zeta\nu(1 + \epsilon \sin \psi\tau)(1 - I_2)I_1 - \zeta I_2$$

[5.6]

where the dimensionless time is $\tau = t/d_1$ and the dimensionless parameters are $\mu = \lambda_{12}d_1/r$, $\psi = wd_1$, $\nu = \lambda_{21}d_2 r$ and $\zeta = d_1/d_2$.

When the equilibrium point [5.2] is translated to the origin by letting $I_1 = E_1 + U$ and $I_2 = E_2 + V$, the model [5.6] becomes

$$\frac{dU}{d\tau} = -(\mu/R)U + RV - \mu UV + \mu \epsilon \sin \psi\tau [R/\mu-U][E_2+V]$$

$$\frac{dV}{d\tau} = (\zeta/R)U - \zeta\nu RV - \zeta\nu UV + \zeta\nu\epsilon \sin \psi\tau [1/\nu R-V][E_1+U]$$

[5.7]

where $E_1 = (\mu\nu-1)/\nu(1+\mu)$, $E_2 = (\mu\nu-1)/\mu(1+\nu)$ and $R = E_1/E_2$. Note that $\mu\nu > 1$ since $K > 1$. It is reasonable to expect that the small periodic forcing in system [5.7] leads to a small periodic solution around the equilibrium point. In fact we show that there is a unique periodic solution which is uniformly asymptotically stable and then we analyze the asymptotic behavior of the first two terms in the power series expansion in ϵ .

System [5.7] is of the form $x' = f(\tau,x,\epsilon)$ where x is a vector function of dimension 2 and f is periodic in τ with period $2\pi/\psi$. The system [5.7] with $\epsilon = 0$ has no nonzero $2\pi/\psi$ periodic solution. By a theorem based on the implicit function theorem (Miller and Michel, 1982, p. 313), for sufficiently small ϵ , system [5.7] has a unique solution $\phi(\tau,\epsilon)$ which is $2\pi/\psi$ periodic and continuous, and such that $\phi(\tau,0)$ is the trivial solution. Since the characteristic roots of the linearization of [5.7] with $\epsilon = 0$ are negative real numbers, the solution $\phi(\tau,\epsilon)$ is uniformly asymptotically stable.

The preceding paragraph shows that a regular perturbation analysis for small ϵ can be used and that there is no danger of parametric resonance or secular terms. In contrast, Dietz (1976) showed numerically that in a measles model with a spiral equilibrium point, seasonal oscillations in the contact rate lead to biennial oscillations in incidence because of subharmonic resonance.

We expand an arbitrary solution in powers of ϵ with the form

$$U(\tau,\epsilon) = U_0(\tau) + \epsilon\, U_1(\tau) + \epsilon^2\, U_2(\tau) + \ldots$$

$$V(\tau,\epsilon) = V_0(\tau) + \epsilon\, V_1(\tau) + \epsilon^2\, V_2(\tau) + \ldots$$

[5.8]

The terms $U_0(\tau)$ and $V_0(\tau)$ satisfy [5.7] with $\epsilon = 0$ so that they are bounded. Since the eigenvalues of the linearization of [5.7] with $\epsilon = 0$ are negative real numbers, $U_0(t)$ and $V_0(t)$ approach the origin exponentially (Miller and Michel, 1982, p. 261).

A straight-forward calculation shows that $U_1(\tau)$ and $V_1(\tau)$ satisfy

$$\frac{dU_1}{d\tau} = -(\mu/R)U_1 + RV_1 - \mu(U_0V_1 + V_0U_1) + \mu\,\sin\,\psi\tau\,[R/\mu - U_0][E_2 + V_0]$$

[5.9]

$$\frac{dV_1}{d\tau} = (\zeta/R)U_1 - \xi\nu RV_1 - \zeta\nu(U_0V_1 + V_0U_1) + \zeta\nu\,\sin\,\psi\tau[1/\nu R - V_0][E_1 + U_0]$$

The matrix form of [5.9] is

$$x'(\tau) = [A + B(\tau)]x + f(\tau) + g(\tau) \qquad [5.10]$$

where

$$x(\tau) = \begin{bmatrix} U_1(\tau) \\ V_1(\tau) \end{bmatrix} \qquad\qquad A = \begin{bmatrix} -\mu/R & R \\ \zeta/R & -\zeta\nu R \end{bmatrix}$$

$$B(\tau) = \begin{bmatrix} -\mu V_0 & -\mu U_0 \\ -\zeta W_0 & -\zeta\nu U_0 \end{bmatrix} \qquad f(\tau) = \begin{bmatrix} E_1\,\sin\,\psi\tau \\ \zeta E_2\sin\,\psi\tau \end{bmatrix}$$

$$g(\tau) = \begin{bmatrix} (RV_0 - \mu E_2 U_0 - \mu U_0 V_0)\,\sin\,\psi\tau \\ (U_0/R - \nu E_1 V_0 - \nu U_0 V_0)\,\zeta\,\sin\,\psi\tau \end{bmatrix}$$

Since $B(\tau)$ and $g(\tau)$ in [5.10] approach zero as τ approaches infinity and the solution of the homogeneous equation is zero, the solution of [5.10] for large time should be determined primarily by $f(\tau)$.

__THEOREM 5.2__ If all eigenvalues of the real constant matrix A have real parts less than $-\sigma$ which is less than 0, B(t) is a real continuous matrix on $[0,\infty)$, f(t) and g(t) are real continuous vector functions on $[0,\infty)$, f(t) is bounded on $[0,\infty)$, $\int_0^\infty B(s)\ ds < \infty$, and $\int_0^\infty g(s)\ ds < \infty$, then solutions of $x'(t) = [A + B(t)]x + f(t) + g(t)$ approach $u(t) = \int_0^t e^{A(t-s)}f(s)ds$ as t approaches ∞.

__PROOF__. Let $y(t) = x(t) - u(t)$. Then

$$y'(t) = [A + B(t)]y + h(t) \tag{P}$$

where $h(t) = B(t)u(t) + g(t)$. Since u(t) is bounded, the finite integral of B(t) implies $\int_0^\infty h(t) < \infty$. To prove the theorem we need only show that __all solutions y(t) of__ (P) __tend to 0 as t $\to \infty$__. This will follow from Strauss and Yorke (1968), Theorem A part ii, in three steps.

First we view the linear system

$$w'(t) = Aw(t) + B(t)w(t) \tag{L}$$

as a perturbed form of $x'(t) = Ax(t)$. Since 0 is (uniform) asymptotically stable, we may apply Theorem A part ii and conclude 0 is "eventually uniform asymptotically stable" (EvUAS) for (L). We refer the reader to the source for a detailed definition, but __all solutions w(t) of__ (L) __tend to 0 as t $\to \infty$__.

Next we view (P) as a perturbed form of (L) and again apply Theorem A part ii to conclude 0 is EvUAS for (P). This implies that __(P) has some solution y_1 satisfying $y_1(t) \to 0$ as t $\to \infty$__.

Finally let $y_2(t)$ be any other solution of (P). Then $w(t) = y_2(t) - y_1(t)$ satisfies (L) and so tends to 0 as t $\to \infty$. Since $y_1(t) \to 0$, we have $y_2(t) \to 0$ as t $\to \infty$. The proof is complete.

All of the assumptions in Theorem 5.1 are satisfied by the $A, B(\tau)$, $f(\tau)$ and $g(\tau)$ in [5.10] since $U_0(\tau)$ and $V_0(\tau)$ are bounded and approach the origin exponentially. The particular solution corresponding to the forcing term $f(\tau)$ can be found by converting the system [5.9] to a second order differential equation. As $\tau \to \infty$ we obtain

$$x(\tau) \to \int_0^\tau e^{A(\tau-s)}f(s)ds \longrightarrow \begin{bmatrix} C_1 \cos\psi\tau + C_2 \sin\psi\tau \\ C_3 \cos\psi\tau + C_4 \sin\psi\tau \end{bmatrix} \tag{5.11}$$

where

$$D = [\zeta(\mu\nu-1) - \psi^2]^2 + (\mu/R + \zeta\nu R)^2\psi^2$$

$$DC_1 = -\psi[\zeta + (\zeta\lambda R)^2 + \zeta(\mu/R + \zeta\nu R) + \psi^2]E_1$$

$$DC_2 = [(1 + \nu R)(\zeta^2)(\mu\nu - 1) + (\mu/R - \zeta)\psi^2]E_1$$

$$DC_3 = -\zeta\psi[\zeta + (\mu/R)^2 + (\mu/R + \zeta\nu R) + \psi^2]E_2$$

$$DC_4 = \zeta[(\mu/R + 1)\zeta(\mu\nu - 1) + (\zeta\nu R - 1)\psi^2]E_2$$

Thus if the relative seasonal change ϵ in the contact rates is small, then solutions of [5.7] approach ϵ times the periodic solution in [5.11] for large time. Solutions (I_1, I_2) of [5.5] approach the equilibrium point (E_1, E_2) plus ϵ times the periodic solution in [5.11] for large time.

We are interested in the size of the oscillations in the prevalences in women and men and in the relationships between the time when the peak contact rate occurs and the times when the peak infective fractions occur. Now $U_1(t)$ has a maximum of $(c_1^2 + c_2^2)^{1/2}$ and a minimum of $-(c_1^2 + c_2^2)^{1/2}$ at times $t = \arctan(c_2/c_1)/w$. The maximum fractional change in the prevalence in women is

$$\frac{\epsilon(c_1^2 + c_2^2)^{1/2}}{E_1} = \epsilon\left(\frac{w^2 + \zeta^2(1+\nu R)^2}{D}\right)^{1/2} \qquad [5.12]$$

The results for men are analogous with the maximum fractional change given by

$$\frac{\epsilon(c_3^2 + c_4^2)^{1/2}}{E_2} = \epsilon\rho\left(\frac{w^2 + (1 + \mu/R)^2}{D}\right)^{1/2} \qquad [5.13]$$

Table 5.3 shows the calculation results for the same sets of parameter values as in Table 5.1. The maximum fractional changes are given by [5.12] and [5.13]. The phase shifts are the number of days that peak prevalence lags behind the peak contact rate. For example, using parameter set 1 a 1% oscillation in the contact rate causes a 1.09% oscillation in the prevalence for women and a 1.37% oscillation in the prevalence for men. The peak prevalence for men occurs 62 days

PHASE PLANE PORTRAIT OF PERIODIC SOLUTIONS CORRESPONDING TO SEASONAL OSCILLATIONS

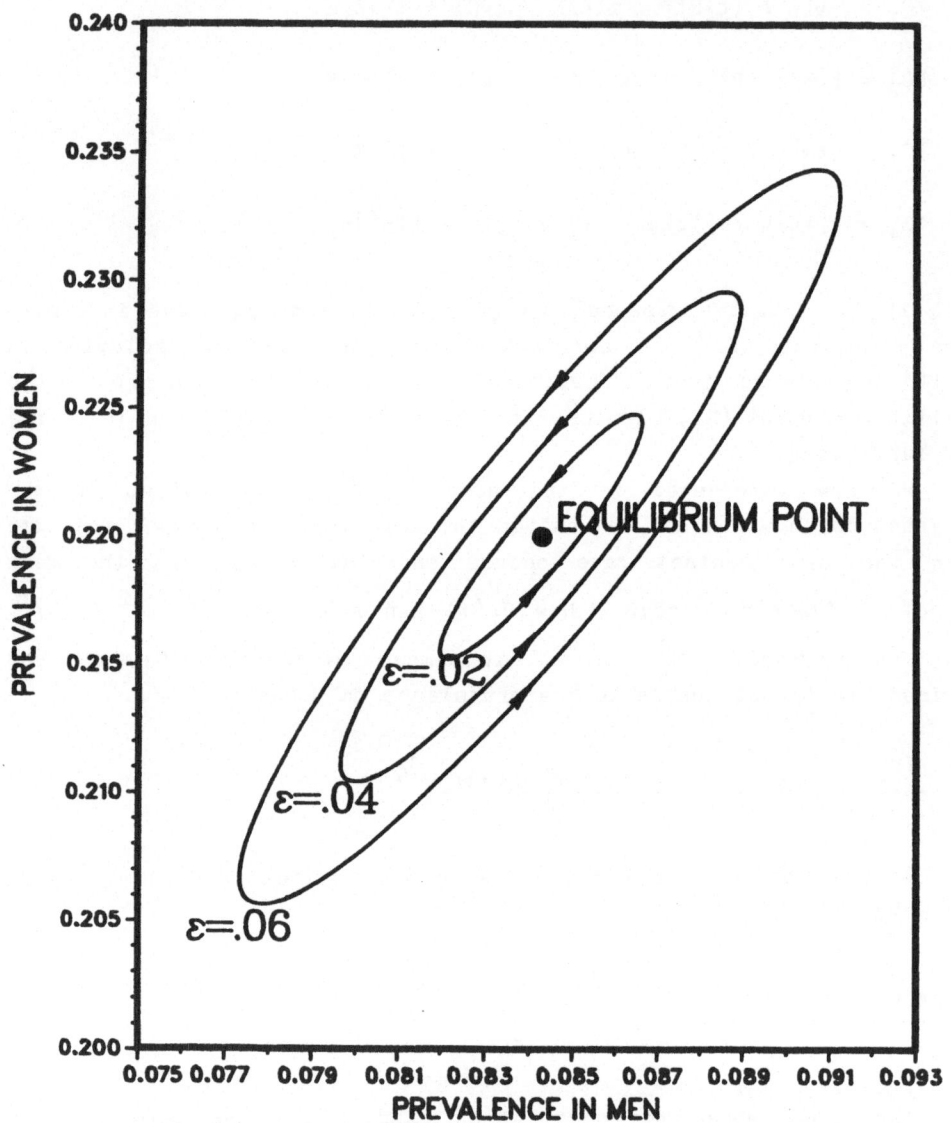

Figure 5.6. Approximate periodic solutions when the contact rate varies seasonally. For a given ε and parameter set 1, solutions of (5.5) approach these periodic solutions.

after the peak in the contact rates and the peak prevalence for women occurs another 22 days later. The pattern is similar for the other parameter sets.

Using parameter set 7, the approximate periodic solution of [5.5] found from [5.8] and [5.11] is

$$I_1(t) = .220 + \varepsilon(-.238 \cos wt + .030 \sin wt)$$

$$[5.14]$$

$$I_2(t) = .084 + \varepsilon(-.101 \cos wt + .056 \sin wt).$$

Figure 5.6 shows approximate periodic solutions around the equilibrium point for small values of ε. The global asymptotic stability mentioned earlier means that for a given ε, all solutions of [5.5] starting with nonzero initial prevalences approach a periodic solution which is closely approximated by the periodic solutions given by [5.14] and shown in figure 5.6.

TABLE 5.3

Amplitude and phase shifts of the forced oscillations
for various parameter sets.

parameter set	1	2	3	4	5	6	7
duration d_1	80	160	160	80	80	80	80
duration d_2	20	40	20	40	20	20	20
contact number K	1.4	1.4	1.4	1.4	2.0	1.2	1.4
parameter er	.5	.5	.5	.5	.5	.5	1.0
ε	.01	.01	.01	.01	.01	.01	.01
oscillation amplitude women	1.09%	0.53%	0.61%	0.90%	0.97%	1.12%	1.09%
oscillation amplitude men	1.37%	0.86%	1.04%	1.07%	1.31%	1.37%	1.34%
phase shift (days) women	84	94	91	86	67	90	84
phase shift (days) men	62	62	50	72	50	67	62

The __maximum__ __fractional__ __change__ in reported incidences is the difference between the maximum and minimum incidences divided by the sum of the maximum and minimum incidences. Seasonal indices (0.942, 0.968, 1.098, 0.989) for all reported cases are given in figure 5.2. Using the seasonal indices for women and men for the years 1964 to 1975 the maximum fractional change in reported incidences are (1.064 − .955)/(1.064 + .954) = .054 for women and (1.107 − .923)/(1.107 +

.923) = .091 for men. These estimates are crude since they are based on quarterly data. The actual oscillations in incidence are probably around 6% in women and 10% in men. It is not possible to estimate the actual phase shifts; however, the quarterly data shows that the peak incidence in women probably occurs about 2 to 3 weeks after the peak incidence in men.

Using parameter set 1 in Table 5.3 it seems that the actual seasonal oscillations in incidence would be caused by a 5% to 7% seasonal oscillation in the contact rate (ε = .05 to .07). Moreover, the actual phase shifts may be about 9 weeks for men and 12 weeks for women. Thus the model suggests that the observed peaks in gonorrhea incidence which occur in August to October are probably due to peak contact rates in June or July.

Thus the first important conclusion is that the observed seasonal oscillations in incidence may be due to reasonably small (5% to 7%) oscillations in the contact rate. The second conclusion is that the observed peak incidences in August to October may be due to a peak contact rate in the summer months. These results were surprising to the epidemiologists in the VD Control Division of the Center for Disease Control when we first announced them.

CHAPTER 6

MODELING GONORRHEA IN A POPULATION
DIVIDED INTO EIGHT GROUPS

In Chapters 4 and 5, the population was divided into two groups. Here the population is divided according to sex, the level of sexual activity and whether the infections are symptomatic or asymptomatic. There are eight groups representing the eight combinations, such as the male-highly active-asymptomatic group. This population dynamics model is used to compare the effectiveness of six control methods for gonorrhea involving population screening and contact tracing of selected groups. A condensed version of the results in this chapter appeared in a paper (Hethcote, Yorke and Nold, 1982).

In section 6.1 the eight groups are described and the system of eight nonlinear ordinary differential equations is given. The number of independent contact rates among people in the eight groups is reduced by using proportionate mixing assumptions in section 6.2. It is shown that a contact number determines whether the disease dies out or remains endemic. The six gonorrhea control methods considered involve screening of women and men, contact tracing women and men who are infectees and contact tracing women and men who are infectors. The six control methods are incorporated into the eight group model in section 6.3.

Estimates of parameters used in the equations are described in section 6.4. Some values of parameters are found from current data and estimates of epidemiologists while other values such as the levels of sexual activity are found indirectly so that incidences and prevalences are consistent with observations and so that the effects of the screening program correspond to observed incidence changes. A computer program finds the endemic equilibrium levels for a given parameter set for each of the six control methods. This program is described in section 6.5; a listing of the program and a sample output are given in appendices. A table summarizing the calculations for six different parameter sets is given in section 6.6.

Since gonococcal infection in a woman can lead to pelvic inflammatory disease and sterility, our criterion for the effectiveness of a control procedure is the extent to which the equilibrium prevalence (and hence the number of months of infection) in women is reduced when the control is added to the equations. The calculations in section 6.6 show that discovering (and curing) **an infectious woman**

by tracing infectors of diagnosed men **is more effective** in this sense **than** discovering (and curing) an infectious woman by **screening,** and **the latter is more effective than** discovering (and curing) an infectious woman by **tracing contacts** of diagnosed men. The calculations also show that the relative effectiveness of using the corresponding procedures to discover infectious men instead of women is approximately the same. Of course, it is more difficult to discover infectious men because male prevalence is much lower.

The control procedures are small supplements added onto the current screening program. The calculations assume that the suplementary control procedures all discover the same number of individuals per unit time, namely, a number equal to 1% of the incidence in women. All calculations are made when the prevalences are at equilibrium. In section 6.7 we will consider conclusions from the model, the relative difficulties of discovering one individual by each of the procedures and implications for gonorrhea control strategies. The results of our theoretical modeling are meant to advise gonorrhea control strategists and clinicians.

6.1 The Model for a Heterogeneous Population

Gonorrhea transmission occurs in a population in which some infectives are more active sexually than others, in which probabilities of transmission per sexual contact are quite different for men and women, and in which some infectives are asymptomatic with long durations of infection while others are symptomatic with shorter durations. A model with eight groups is needed to incorporate all of these essential aspects. Although sexually transmitted diseases are a major health problem among homosexuals, there does not seem to be much transmission between the homosexual population and the heterosexual population. Consequently, we consider a heterosexual population, i.e., we assume that all contacts are heterosexual. The population considered here is the sexually active women who are the target of the screening program in the United States and their male partners.

The four groups of women have odd indices $(1,3,5,7)$ and the four groups of men have even indices $(2,4,6,8)$. Let N_i denote the size of the group i. Since the sizes of the odd (even) index groups can be divided by the total number of women (men) in the population, we assume that N_i is the fraction of women or men in group i. Hence

$$N_1 + N_3 + N_5 + N_7 = 1 \text{ and } N_2 + N_4 + N_6 + N_8 = 1.$$

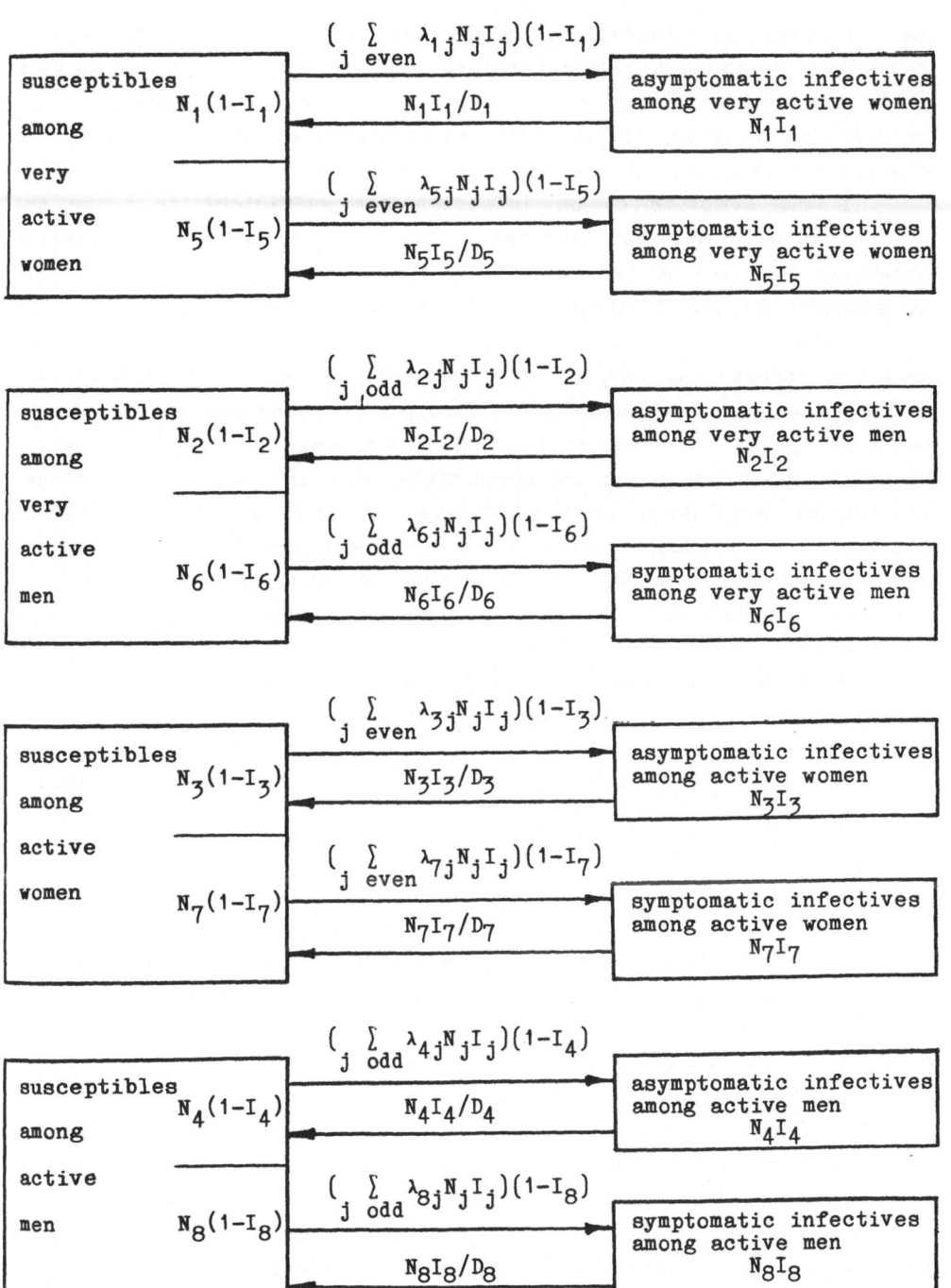

Figure 6.1.　Flow diagram for the eight group model.

As indicated in Chapter 1, gonococcal infection does not confer resistance or immunity against reinfection so that individuals in each group are either susceptible or infectious. As in section 3.1, the prevalence in group i at time t (in months) is $I_i(t)$ and the susceptible fraction of the population is $1-I_i(t)$.

If the **symptoms** of an infection **are sufficient to cause the person to seek medical treatment**, then the person is **symptomatic**; otherwise, the person is asymptomatic. As before, the sexual activity is measured by the frequency of encounters (cf. section 3.2). Group 1(2) consists of very active women (men) who are asymptomatic when they are infectious. Group 3 (4) consists of active women (men) who are asymptomatic when infectious. Group 5 (6) consists of very active women (men) who are symptomatic when infectious. Group 7 (8) consists of active women (men) who are symptomatic when infectious. Of course, the actual population is very diverse so that it does not divide clearly into groups; however, it is convenient conceptually and computationally to assume the existence of these groups so that the effectiveness of various control procedures can be evaluated and compared.

The model consists of the differential equations and initial conditions presented in section 3.1 with n=8.

$$\frac{d}{dt}(N_i I_i) = \sum_{j=1}^{8} \lambda_{ij}(1-I_i)N_j I_j - N_i I_i/D_i \qquad [6.1]$$

$$I_i(0) = I_{io} \qquad i = 1,2,\ldots,8. \qquad [6.2]$$

Note that here $\lambda_{ij} = 0$ if i+j is even since there are only heterosexual contacts in the model. We assume that the λ_{ij} are fixed and do not vary seasonally. Here we use one month as the unit time so D_i is the mean duration of infection in months. A flow diagram for this model is given in figure 6.1. Modifications of the equations [6.1] to include control procedures will be described in section 6.3.

The first challenge in dealing with a model of this complexity is choosing the parameters in a simple and rational way, in a way sufficiently clear that the reader can feel confident that a reasonable collection of parameter choices has been tested. There are, after all, thirty two nonzero λ_{ij}s, eight N_is and eight D_is. Reducing the number of parameters will be a major concern in this chapter.

In the model, we have separated women into groups according to whether or not their infections are asymptomatic. Wiesner and

Thompson (1980) state that the infected individual's ability to recognize symptoms of gonococcal infection and then get medical treatment may vary significantly from person to person. However, it is also reasonable to assume that very active women are distributed between being asymptomatic and symptomatic when infectious in a random manner. Our model is consistent with this assumption if we assume that the susceptibles among very active women are sometimes asymptomatic and sometimes symptomatic. Similar remarks also apply to the other groups. Other models involving asymptomatics have also been considered (Kemper, 1978; Bailey, 1979; Cooke, 1982).

As in the earlier models we will consider the positive equilibrium point and how it changes when parameter values change or when control procedures are added. Let E_i be the endemic equilibrium prevalence in group i. Then the E_i are the solutions of the 8 simultaneous quadratic equations obtained by setting the right sides of [6.1] equal to zero. Since the first term in each differential equation corresponds to the incidence (number of new cases per unit time), the equilibrium incidence in group i is equal to the equilibrium prevalence E_i times the group size N_i divided by the mean duration D_i (cf. section 2.1).

6.2 The Contact Matrix

The contact matrix has 32 zero entries and 32 positive entries which must be determined. As in section 3.2, we will initially use a proportionate mixing approach to determine these contact rates, i.e., we assume that the number of encounters between groups of women and men is proportional to the relative sexual activities of the groups.

Let A_j be the activity level of group j which is the average number of encounters (with different partners) of a person in group j per unit time. Let Q_j be the probability that an infective in group j transmits the disease during an encounter with a susceptible, i.e., that there is an adequate contact. Let M_{ij} be the fraction of encounters made by an average infective of group j with persons in group i. Hence $\Sigma_i M_{ij} = 1$ for each j. The matrix M is called the mixing matrix. Using these definitions it follows that the average number of encounters per unit time of an infective in group j with different partners in group i is $A_j M_{ij} = \lambda_{ij}/Q_j$.

The fact that each sexual encounter involves one man and one woman requires that the average number of encounters per unit time for women (sum of $A_i N_i$ over odd indices) must equal the average for men (sum of $A_i N_i$ over even indices); we denote this value by A. We define the activity fraction B_i of groups i by $B_i = A_i N_i / A$. Hence

$$B_1 + B_3 + B_5 + B_7 = 1 \quad \text{and} \quad B_2 + B_4 + B_6 + B_8 = 1.$$

We assume all $A_i > 0$ so that all $B_i > 0$. The proportionate mixing assumption is that the encounters of a person are distributed in proportion to the activity fractions of the groups of the opposite sex, i.e., $M_{ij} = B_i$ for all i and j such that i + j is odd. Note that the proportionate mixing formulation here is different from that in section 3.2 since there are groups of women and men here with heterosexual contacts.

We let $K_j > 0$ denote the contact number for group j, which is the number of adequate contacts made by a typical infective of group j during the duration of infection so that $K_j = Q_j A_j D_j$. The number of adequate contacts T_{ij} of a group j infective with group i during an average case satisfies $T_{ij} = \lambda_{ij} D_j = A_j M_{ij} Q_j D_j = M_{ij} K_j$ We call the matrix $[T_{ij}]$ the transmission matrix. **Proportionate mixing means that**

$$T_{ij} = B_i K_j$$

for i+j odd and $T_{ij} = 0$ for i+j even. Notice that the matrix $[T_{ij}]$ is determined by 16 values, the B_is and K_js.

The second generation contact number for this model with proportionate mixing is

$$K = (\sum_{i \text{ odd}} B_i K_i)(\sum_{j \text{ even}} B_j K_j),$$

where the first factor is the average number of men adequately contacted by an average infectious woman during her infectious period, and the second factor is the average number of women adequately contacted by an average infectious man. Thus K is the number of women contacted in two generations, that is, by an infectious woman through her contacts with men (first generation) and their contacts with other women (second generation). The theorem below means that **this second generation contact number is a threshold parameter which determines whether the disease dies out** (K < 1) **or remains endemic** (K > 1). The second generation contact number has the same intuitive interpretation as in section 5.1.

The characteristic equation for the transmission matrix $[T_{ij}]$ is $\det([T_{ij}] - \alpha I) = \alpha^6(\alpha^2 - K) = 0$. This equation follows from a detailed calculation using properties of determinants. Notice that $[T_{ij}]$ is irreducible since all B_is and K_js are positive. Recall that irreducibility implies that the whole population cannot be split into two subpopulations which do not interact with each other.

THEOREM 6.1 In the proportionate mixing model, the solutions of equations [6.1] approach the origin if $K < 1$ and they approach a unique positive equilibrium if $K > 1$.

PROOF. From the characteristic equation and Lemma 3.2, the Perron eigenvalue $p(T) = K^{1/2}$ is equal to the spectral radius $r(T)$. By Lemma 3.3, $r(T) = K^{1/2} < 1$ is equivalent to the outbreak eigenvalue satisfying $m_0 = s(V) < 0$. The theorem now follows from Theorem 3.1.

The proportionate mixing assumption is not always completely reasonable, since a very active person may be more likely to have an encounter with a very active person. Very active people may know how to seek very active people. In the extreme case where very active (active) people only have encounters with very active (active) people, then the encounters of a person are distributed in proportion to the fractional activity levels of the opposite sex group with the same activity level. Thus there is proportionate mixing within the very active subpopulation and within the active subpopulation, but there is no interaction between these subpopulations. The mixing matrix for this model has 48 zero entries and 16 positive entries.

Since the actual mixing is probably somewhere between the extremes of proportionate mixing in the entire population and proportionate mixing in the activity levels, we use a mixing matrix M which is 1-G times the mixing matrix for proportionate mixing plus G times the mixing matrix for proportionate mixing in the activity levels. The fraction G between 0 and 1 is called the selectivity constant.

6.3 Six Control Methods

The reduction of pelvic inflammatory disease (PID) is a primary goal of gonorrhea control activities. Some infected women develop PID; consequently, it is reasonable to assume that a reduction in the months of infection of women in the population would cause a corresponding reduction in PID. Therefore **our criterion for the effectiveness of a control procedure is the percentage reduction in prevalence for women.** This is equivalent to measuring the percentage reduction in total months of infection for women each year. Reduction in incidence is a less useful criterion, because control procedures can cure people and make them available for new infections, so that two cases might occur where there would otherwise have been one case of long duration.

The six control methods are screening and targeted contact tracing procedures which are designed to discover infectious women and men so that they can be cured. Let C be proportional to the number of women being screened per unit time. The R_i are the relative rates at which women are discovered and cured by screening in the various groups. C is an adjustable parameter measuring the effort put into the screening program so that CR_i is the rate at which women are discovered and cured in group i. For each of the six supplementary control procedures, there are analogous terms that we denote by E and P_i. Equations [6.1] are modified to include C and E as follows:

$$\frac{d}{dt}(N_i I_i) = \sum_{j=1}^{8} \lambda_{ij}(1-I_i)N_j I_j - N_i I_i/D_i - CR_i - EP_i \qquad [6.3]$$

Our objective is to compare the effects of the six control methods when they are implemented at low levels as supplements to the existing screening program. To be comparable we choose the value of E for each method so that the discoveries are 1% of the incidence in women, i.e.,

$$\sum_{i=1}^{8} EP_i = (.01) \sum_{i \text{ odd}} \sum_{j=1}^{8} \lambda_{ij}(1-I_i)N_j I_j .$$

The six control methods are described below.

Type 1W: Screening of Women

Screening of women consists of culture testing of women at certain health facilities. Screening of women is currently the primary control method in the United States. The number of infectious women discovered by screening is proportional to the prevalence. If the prevalence is doubled in a group, then twice as many will be discovered by screening. Thus

$$R_i = I_i N_i \qquad \text{for i odd}$$

and $R_i = 0$ for i even. Since the type 1W supplementary program also corresponds to screening, $P_i = I_i N_i$ for i odd and $P_i = 0$ for i even.

Type 1M: Screening of Men

This procedure is analogous to type 1W with roles reversed so that $P_i = I_i N_i$ for i even and $P_i = 0$ for i odd. (We remark that screening of men produces so few discoveries that it is generally impractical.)

Type 2W: Contact Tracing Women Who Are Infectees

An <u>infectee</u> is a person to whom the gonococcal infection has been spread by the reference case (the patient who has come in for treatment) and the <u>infector</u> is the source of the gonococcal infection in the reference case.   The supplementary control procedure 2W consists of discovering women infectees by tracing and culture testing women who are named as contacts (but not as the infector) by diagnosed men.   Infected men of group j are diagnosed (and cured) at rate $I_j N_j / D_j$ and the average man in group j during the duration of his infection infects $\lambda_{ij}(1-I_i)D_j$ women in group i.   Therefore, the number of female infectees per unit time in group i is

$$P_i = \sum_{j \text{ even}} \lambda_{ij}(1-I_i)I_j N_j \text{ for i odd}$$

and $P_i = 0$ for i even.   Hence type 2W discoveries are distributed in proportion to the rate at which new cases occur in the female groups, i.e., in proportion to the incidence.

Type 2M: Contact Tracing Men Who Are Infectees

This procedure is analogous to type 2W with roles reversed so that $P_i = \sum_{j \text{ odd}} \lambda_{ij}(1-I_i)I_j N_j$ for i even and $P_i = 0$ for i odd.   Note that 2W and 2M involve tracing contacts and do not include tracing the named infector or source of the infection.

Type 3W: Contact Tracing Women Who Are Infectors

In this procedure, infectious women are identified through the men they infect.   The probability that a diagnosed man has been infected by a woman in group i is proportional to the rate at which infections are caused by women in group i.   Hence the discoveries are distributed using

$$P_i = \sum_{j \text{ even}} N_i I_i \lambda_{ji}(1-I_j) \quad \text{for i odd}$$

and $P_i = 0$ for i even.

Type 3M: Contact Tracing Men Who Are Infectors

In this procedure, the discoveries are distributed using

$$P_i = \sum_{j \text{ odd}} N_i I_i \lambda_{ji}(1-I_j)$$

for i even and $P_i = 0$ for i odd.

Screening and control types 1W and 1M change the usual removal terms in [6.3] so that [6.3] is like [3.1], but with modified durations. Control types 2W and 2M change the incidence terms in [6.3] so that [6.3] is like [3.1] when the terms in [6.3] are combined, but with a different contact matrix. The signs of the entries are unchanged for small supplementary control efforts. Control types 3W and 3M modify the linear and quadratic terms in [6.3], but for small E [6.3] is like [3.1] with a different contact matrix and different durations. Since the models with controls are forms of [3.1], they are covered by Theorem 3.1. Thus we can examine the coordinates Y_i of the endemic equilibrium point corresponding to specific epidemiologic and control parameters. As indicated in the introduction to this chapter, control procedures are compared by comparing their effect on the equilibrium prevalence in women.

6.4 Parameter Estimation

The following three criteria are used as reasonable restrictions on the parameter sets for the model:

1. the incidence for men must be slightly larger than the incidence for women;

2. the prevalence for women must be approximately 3% of the population of women at risk;

3. discovering 10% of the infective women by population screening must result in a 20% decrease in incidence in men.

The first criterion is based on observed gonorrhea incidence (Wiesner and Thompson, 1980; CDC, 1981a) The second criterion is used because prevalences exceeding 3% seem unrealistic as discussed in section 4.1. Yorke, Hethcote and Nold (1978) estimated using a trend analysis that approximately 10% of all actual cases of gonorrhea in women in the United States were discoveries of the culture-screening program. They also estimated that the result of this discovery of 10% of infectious women was an approximately 20% decrease in actual incidence in men. The third criterion corresponds to these estimates.

The system [6.3] contains 48 parameters: 32 positive λ_{ij}, 8 D_i and 8 N_i. Under some simplifying assumptions described below, the 48 parameters can be reduced to the following 8: the duration of infection for all asymptomatics (D_a), the duration of infection for all symptomatics (D_s), the fraction of men who are asymptomatic when infectious (A_m), the fraction of women who are asymptomatic when infectious (A_w), the ratio (J) of the transmission probabilities during an encounter by an infectious man and by an infectious woman, the fraction (F) of the women or men who are in the very active groups, the ratio (H) of the sexual activity of the very active to the active persons, and the selectivity constant (G). These constants and estimates for them are discussed below.

We assume that there is one typical mean duration for the four groups of asymptomatics and another for the four groups of symptomatics. The duration ranges for asymptomatic women and men are estimated to be 3-12 months and 3-6 months, respectively (Wiesner and Thompson, 1980). The duration ranges for symptomatics are estimated to be 3-45 days for women and 2-30 days for men (Wiesner and Thompson, 1980). Hence we use D_a = 6 months and D_s = .25 or .5 month. The results depend primarily on the ratio D_a/D_s so that halving or doubling both D_a and D_s has only a slight effect on the relative merits of the control procedures. We assume that the fractions which are asymptotic when infectious are A_w = .6 for women and A_m = .1 for men. These parameter values lead to results which are consistent with estimates in Wiesner and Thompson (1980) that: 30-60% of the incidence in women are asymptomatics; asymptomatics are 80-98% of the prevalence in women and, consequently, are responsible for 80-98% of the transmissions; 2-5% of the incidence in men are asymptomatics and asymptomatic men account for 60-80% of the prevalence in men and hence 60-80% of the transmissions.

During sexual intercourse the probability of transmission from an infectious woman to a susceptible man is estimated to be .2 to .3 while the probability of transmission from an infectious man is approximately .5 to .7 (Wiesner and Thompson, 1980). Because an encounter involves one or more sexual intercourses, the transmission probability for women might be .5 while the transmission probability for men could be .9. The model requires an estimate of the ratio J of the transmission probability for men to that for women. This ratio J is taken to be 2 or 1.

Assume that the fraction F who are in the very active groups and the ratio H of sexual activity of the very active to the active persons is the same for women and men. In our model, we assume that 1 to 3% of the population is very active and that very active people are 5 to 10 times as sexually active as active people. The population size fractions N_i can be calculated by using N_1 = $(F)(A_w)$, N_3 = $(1-F)(A_w)$, N_5 = $F(1-A_w)$, N_7 = $(1-F)(1-A_w)$ and analogous formulas for N_2, N_4, N_6 and N_8.

The contact numbers K_i (relative to K_1) are found from the durations D_i, the population sizes N_i, the transmission probability ratio J and the activity level ratio H by using

$$K_2 = \frac{K_1 D_2 J}{D_1} (N_1 + N_5 + \frac{N_3 + N_7}{H})/(N_2 + N_6 + \frac{N_4 + N_8}{H})$$

$K_3 = K_1 D_3/(D_1 H)$, $K_4 = K_2 D_4/(D_2 H)$, $K_5 = K_1 D_5/(D_1)$, $K_6 = K_2 D_6/(D_2)$, $K_7 = K_1 D_7/(D_1 H)$, and $K_8 = K_2 D_8/(D_2 H)$. These relationships are derived by using $K_i = Q_i A_i D_i$, $A_1 = HA_3 = A_5 = HA_7$, $A_2 = HA_4 = A_6 = HA_8$, and the conservation of encounters. The contact number K_1 determines the absolute level of sexual contacts and is chosen so that the third criterion is satisfied as explained in the next paragraph. The contact numbers and the durations are used to determine the activity levels. These activity levels and the population size fractions are used to determine the mixing matrices for the proportionate mixing model and for the model with proportionate mixing within activity levels. The selectivity constant G which combines these two mixing matrices is chosen so that the second criterion is satisfied. If G is zero, prevalences in the model are usually unrealistically high. The selectivity constant G measures the correlation between the activity level of the infectious person and the activity level of the sexual partner. The correlation coefficient can be shown to be

$$r = G[(\frac{1-G}{1+FH-F} + G) (\frac{H(1-G)}{1+FH-F} + G)]^{-1/2}$$

In the proportionate mixing model (when G is zero), the correlation coefficient r is zero.

The constant C in [6.3] is adjusted so that 10% of the equilibrium incidence in women is discovered by population screening. The third criterion requires that the absolute level of sexual contacts is adjusted so that the result is a 20% decrease in incidence in men. If the absolute level were too low, then the prevalences would be zero or the addition of C could be sufficient to drive the prevalences to zero. If the absolute level were too high, then the addition of population screening would cause only a small change in incidence in men. The correct absolute level of sexual contacts is determined by an iterative process using a separate computer program.

The extra cure rate is supplementary since it is a small control effort added onto the existing population screening program. For the supplementary control procedures, E is chosen so that the number of extra discoveries are equal to 1% of the equilibrium incidence in women. Hence the supplementary control procedures are directly comparable since they all involve the same number of discoveries.

6.5 The Computer Programs

Computer programs have been written to find the endemic equilibrium point from the simultaneous quadratic equations obtained by setting the right sides of [6.3] equal to zero. The input requested by the programs are values related to the eight essential parameters described in the previous section. The programs first construct the contact matrix $[\lambda_{ij}]$, and then solve the differential equations [6.3] numerically using Euler's method until the prevalences are near the equilibrium levels. Finally these good approximations are used as starting values in Newton's method to find the coordinates of the equilibrium point. Although a model with 8 groups is considered here, similar computer programs have been used to compare control procedures for models with 4 and 12 groups.

The computer program FINDK uses an iterative procedure to find the value of the contact number K_1 of the first group so that if the number of women discovered by general population screening is 10% of the incidence in women, then there is a 20% decrease in the incidence in men. The computer program GCCONT requests input parameter values and then produces output tables consisting of prevalences, incidences, percent changes and preventions for five cases: the model with no control procedure, the model with general population screening and then the model with general population screening plus one of the three supplementary control procedures 1W, 2W or 3W for women. The computer program SCRMEN is similar except that it produces output for the 3 supplementary control procedures 1M, 2M or 3M for men. Appendix 1 contains the listing of the computer program GCCONT. This program contains many remark statements which explain the steps in the program. The programs FINDK and SCRMEN are modifications of the GCCONT. These computer programs were written in the BASIC language by Annett Nold and Herb Hethcote.

Appendix 2 contains a sample run of GCCONT showing input parameter values and output tables. This sample run produces the data for 1W, 2W and 3W in Table 6.1 corresponding to parameter set 1. Note that $N_1 + N_3 = .60$ and $N_2 + N_4 = .10$ so that .6 of the women and .1 of the men are asymptomatic when infectious. Since $N_1 + N_5 = .02$ and $N_2 + N_6 = .02$, 2% of the women and men are very active. The average durations are $D_a = 6$ months for asymptomatics and $D_s = .5$ months for symptomatics. The sexually active groups are 10 times as active as the less active groups, the transmission probability ratio is 2 and the selectivity constant is .2. Notice that the three criteria given in the previous section are satisfied. In particular note that the value K_1 = 9.238 determined by FINDK causes a 20.0018% decrease in the monthly incidence in men due to general population screening of women. Although the output tables contain detailed information, the sentences at the bottom of each output table contain the most useful information. The percentage decreases and cases prevented corresponding to 1W, 2W and 3W for parameter set 1 were obtained from these sentences.

6.6 Comparison of the Supplementary Control Procedures

In Table 6.1 we present the results of calculations with six different parameter sets. Six of the parameters are somewhat arbitrary, namely, the fraction F of the population in the very active groups, the selectivity constant G, the ratio H of sexual activity of very active persons to active persons, the duration D_a of asymptomatics, the duration D_s of symptomatics and the transmission probability ratio J. Therefore we compute the equilibria with several sets of values of these parameters. Thus the parameters F, G, H, J, D_a, D_s take on a variety of values, while the other essential parameters are fixed at the values given in section 6.4. The three criteria given in section 6.4 for a parameter set to be reasonable are satisfied for these parameter sets.

As described earlier, we use the percentage decrease in the population prevalence in women as the measure of the effectiveness of a supplementary control procedure. In Table 1 the percentages can be compared directly since each of the six supplementary procedures results in the same number of discoveries. For parameter set number 1, type 3W control (supplementary tracing of infectors of diagnosed men) is 1.9 times as effective per discovery in reducing prevalence as type 1W control (supplementary population screening) and is 2.8 times as effective per discovery as type 2W control (supplementary tracing of contactees of diagnosed men). Control procedure type 3M is 2.2 times as effective per discovery as type 1M and is 5.4 times as effective per discovery as type 2M. Types 1M and 3M are slightly more effective than 1W and 3W, respectively, while 2M is less effective than 2W.

Another measure of the effectiveness of a supplementary control procedure is the number of cases in women and men prevented by the discovery and cure of one infectious person through this procedure. To evaluate the number prevented, the new equilibrium is found when a certain number of cases per day are cured. Then we can determine how much incidence has dropped, that is, how many fewer women and men are

TABLE 6.1

Summary of Output From the Computer Program
for Various Input Parameter Sets

Parameter Set Number		1	2	3	4	5	6
F = fraction in very active groups		.02	.01	.03	.02	.02	.02
H = sexual activity ratio		10.	10.	5.	10.	10.	10.
G = selectivity constant		.2	.2	.5	.2	.2	.2
r = correlation coefficient		.08	.08	.31	.08	.08	.08
D_a = duration of asymptomatics		6.	6.	6.	6.	6.	12.
D_s = duration of symptomatics		.5	.5	.5	.25	.5	.5
J = transmission probability ratio		2.	2.	2.	2.	1.	2.
population prevalence with { women		.028	.031	.027	.028	.026	.028
general population screening { men		.011	.012	.011	.010	.015	.010
population incidence with { women		.009	.010	.009	.010	.008	.005
general population screening { men		.011	.012	.012	.014	.015	.07
percentage decrease in prevalence in women due to supplementary control procedure	1W	4.4	4.2	4.1	4.3	4.3	4.3
	2W	3.0	2.9	2.7	2.7	2.9	2.7
	3W	8.4	8.1	5.5	7.9	8.3	7.9
	1M	4.8	4.7	4.5	5.9	2.9	5.9
	2M	2.0	2.0	1.8	1.6	1.3	1.6
	3M	10.7	9.8	6.3	12.9	6.1	12.9
cases in women and men prevented per person and cured by the supplementary control procedures	1W	6.5	6.0	6.1	6.8	8.3	6.8
	2W	4.5	4.2	4.2	4.4	5.7	4.4
	3W	16.3	15.3	9.9	16.7	20.6	16.7
	1M	8.7	8.2	8.1	11.4	6.4	11.4
	2M	3.8	3.6	3.5	3.3	3.0	3.3
	3M	19.9	17.9	11.9	25.7	13.9	25.7

being infected each day. Using parameter set number 1, the cases in women and men prevented by the discovery of one infectious woman by type 3W are 2.5 times the preventions by type 1W and 3.6 times the preventions by type 2W control. The cases prevented by the discovery of one infectious man by type 3M control are 2.3 times the preventions by type 1M and 5.2 times the preventions by type 2M. Thus the control procedures have the same relative merits using this measure of effectiveness.

The heterogeneity of the population is a result of the sexual activity ratio H, the ratio D_a/D_s of durations of asymptomatics and symptomatics, and the relative likelihood A_w/A_m of a case being asymptomatic. Since parameter sets 2 and 5 describe populations which are approximately as heterogeneous as that of parameter set 1, the percentages are approximately the same. Since parameter set 3 describes a less heterogeneous population than parameter set 1, the percentage decreases in prevalence in women are lower and the ratios are lower than for parameter set 1. Parameter sets 4 and 6 describe more heterogeneous populations than parameter set 1 and the ratios of the percentage decreases in prevalence in women are greater for 3M, 1M and 2M. In a homogeneous population where all people have the same sexual activity levels (H = 1) and the same durations ($D_a = D_s$), the procedures 1W, 2W and 3W are equally effective per discovery and the methods 1M, 2M and 3M are equally effective. From the data in Table 1 and other calculations, we conclude that the more heterogeneous the populations, the larger the ratio of the effectiveness of the control procedures.

Types 3W and 3M are more effective because they have a high likelihood of identifying very active people and a high likelihood of identifying asymptomatics. Although types 1W and 1M tend to identify aymptomatics, they are not very effective in finding very active people. Control types 2W and 2M are not very effective in finding either very active or asymptomatic persons. For example, for parameter set 1, the probability that a discovery is a very active person is .20, .23 and .71 for types 1W, 2W and 3W, respectively; and the probability that a discovery is asymptomatic is .93, .57 and .91 for types 1W, 2W and 3W, respectively.

In section 6.4 we assumed that .6 of the women and .1 of the men are in groups where individuals are asymptomatic when infectious. For parameter set 1, asymptomatic women are 57% of the incidence and are 93% of the prevalence; asymptomatic men are 8% of the incidence and are 63% of the prevalence. Hence asymptomatic women (men) account for

93% (53%) of the transmissions. Although the percentages for asymptomatic men are not quite in the estimated ranges specified in section 6.4 they are closer to the ranges for this parameter set than for the many other parameter sets attempted. Hence these values justify a posteriori the above choices of these two parameter values.

6.7 Discussion

The eight group model is more realistic than the two group models considered in earlier chapters. The comparison of the strategies studied in this chapter seemed to require a model in which all eight groups interact. Use of the eight group model reveals that tracing infectors is approximately three times as effective as tracing infectees. This result is more realistic than that obtained in section 4.4 using the two group model.

Earlier we defined the core as the union of those groups whose prevalence exceeds 20%. From the calculations for parameter set 1 with screening in Appendix 2, we see that the prevalences of the very active asymptomatic groups are 41% for women and 50% for men. These prevalences are much higher than for the other six groups. Thus the model suggests that the core should consist of very active women and men who are asymptomatic when infectious. However, since it is not understood why some cases are asymptomatic and others are symptomatic, it is also useful to calculate prevalences when asymptomatics and symptomatics are merged. The epidemiologist who is trying to identify the core would not be able to tell whether a particular person is potentially symptomatic or asymptomatic and, consequently, would look at these merged populations. By using the data in Appendix 2, we find that the prevalence for all very active women is 27% and the prevalence for all very active men is 12%. **From this vantage point, the core would consist of all very active women.**

Results which are consistent over a range of acceptable parameter values can be considered to be robust predictions of the model. An example of a robust result is that the relationship between the effectivenesses of the supplementary control procedures is always the same for this model; specifically, infector tracing (type 3W or 3M) is more effective per discovery than population screening (type 1W or 1M) and population screening is more effective per discovery than infectee tracing (type 2W or 2M). Although the magnitude of the ratios of the effectivenesses are not robust since they depend on the particular parameter set, we observe that the greater the heterogeneity of the population in terms of the differences in sexual activity and the differences in durations between asymptomatics and symptomatics, the larger the ratios of the effectivenesses of the control types.

The relationship between the effectiveness of the supplementary control types is explained as follows. Asymptomatics are more important transmitters than symptomatics because they are infectious longer. Very active persons are more important transmitters because they contact more people while they are infectious. Infector tracing (types 3W and 3M) is the most effective control procedure because it is more likely to identify both asymptomatics and very active people; population screening (types 1W and 1M) is next in effectiveness because it is more likely to identify asymptomatics; and infectee tracing (types 2W and 2M) is the least effective control procedure per discovery because it is not more likely to identify either asymptomatics or very active people.

The discoveries by the supplementary control procedures are all set equal to 1% of the incidence in women so that all of them are comparable. Since there are approximately 1.0 million women infected each year, each supplementary control procedure would have to identify and cure an extra 10,000 people. We estimate that type 1W would require screening at least an additional 500,000 women per year (Yorke, Hethcote and Nold, 1978). Control types 2W and 3W would require testing contacts of the approximately 600,000 infectious men reported annually. It would therefore be necessary to identify and cure one extra infectious contact for each 60 reported men for type 2W control or one more infector for each 60 reported men for type 3W control. Since about 400,000 women are reported annually, it would be necessary to identify and cure one extra male infectee for each 40 reported women for type 2M control or one more male infector for each 40 reported women for type 3W control.

In constructing the model it is assumed that infectors can be clearly distinguished from contacts so that type 3W is distinct from type 2W and type 3M is distinct from type 2M. Even though identifications of infectors versus contactees will not always be accurate, we still feel that concentrating on tracing people named as infectors is a reasonable practical goal. If only the most recent contacts of the reference cases are traced, then the infectors might be consistently missed. A control program which takes advantage of the findings here is one based on interviewing, in which the interviewer raises the question about the infector and whether this person might be brought in for treatment. Field studies could give valuable ideas on conducting interviews to identify infectors.

Tracing persons named as infectors would sometimes lead to infectees; however, the program would still be more effective than a program which puts no emphasis on finding infectors. The program proposed is modest in scale since it only requires that one extra infector be identified for every 40-60 cases of the opposite sex.

The relative merits of the various control procedures depend not only on the effectiveness per discovery in reducing prevalence, but also on the costs of discovering an infective by the procedures. Some investigators believe that infectives can often identify their infectors. Hence the person named as the infector is often the actual infector and this person (especially an asymptomatic) is often still infectious when contacted and checked. It might be necessary to contact and check more than one person identified as a contact (not the infector) of the reference case in order to find another infective. Since only a few people would have to be contacted, the cost of a discovery and cure by tracing infectors and infectees might be low. Yorke, Hethcote and Nold (1978) estimated that about fifty women must be checked by general population screening (type 1W) in order to discover one infectious woman (excluding people who would be identified without culture screening). However, an add-on culture test for someone who is already being examined is substantially cheaper than the cost of tracing and examining a suspected infector or infectee. Since the prevalence in men is very low, the cost of discovering an infectious man by population screening could be so high that type 1M control is impractical. The results of our analysis of this model are meant to advise gonorrhea control strategists and clinicians.

If the cost in dollars or the relative costs could be estimated for the discovery and cure of one person by each of the procedures, then the cost per prevention of a new case could be computed for each of the control procedures. Comparison of the costs per prevention would show which of the six procedures is the most efficient use of the resources available for gonorrhea control. Different cities or states may find that different control procedures are more effective because of the special characteristics of their population.

LISTING OF THE COMPUTER PROGRAM GCCONT

```
00100  REM     THIS PROGRAM, GCCONT, FINDS PREVALENCE AND INCIDENCE OF GONORRHEA
00120  REM  FOR SEVERAL CONTROL SITUATIONS.TO MAKE A RUN, ENTER PARAMETERS WHEN
00140  REM  THEY ARE REQUESTED.
00180  REM
00200  REM     THE MODEL EQUATIONS ARE COMPUTED IN THE 'RATE' SUBROUTINE AND THE
00220  REM  'CONTROL' SUBROUTINE, USING THE CONTACT MATRIX WHICH THE 'PATTERN'
00240  REM  SUBROUTINE CALCULATED.
00260  REM
00280  REM      THE PROGRAM IS ORGANIZED AS FOLLOWS:
00300  REM  AFTER WE ENTER DATA,THE PROGRAM MAKES PRELIMINARY CALCULATIONS WHICH
00320  REM  INCLUDE CALLING THE PATTERN SUBROUTINE TO FORM THE CONTACT MATRIX.
00340  REM  THEN THE MAIN PROGRAM LOOP BEGINS:DURING A RUN THE LOOP VARIABLE R1
00360  REM  STEPS THROUGH THE FOLLOWING VALUES:
00380  REM      R1=1 FOR THE UNCONTROLLED CASE
00400  REM       R1=2 FOR POPULATION SCREENING (WOMEN)
00420  REM  R1=3 FOR POP SCREENING PLUS SUPPLEMENTARY POP SCREENING
00440  REM  R1=4 FOR POP SCREENING PLUS SUPPLEMENTARY INFECTEE TRACING
00460  REM  R1=5 FOR POP SCREENING PLUS SUPPLEMENTARY INFECTOR TRACING
00480  REM
00500  REM  FOR EACH CASE, THE MAIN PROGRAM CALLS THE 'SOLVER' SUBROUTINE WHICH
00520  REM  USES ITERATION PROCEDURES TO FIND PREVALENCES. THE MAIN PROGRAM ALSO
00540  REM  MAKES SUMMARY CALCULATIONS FOR EACH CASE AND PRINTS OUT A TABLE OF
00560  REM  RESULTS.  THE 'SOLVER' SUBROUTINE ORGANIZES THE ITERATION PROCESS.
00580  REM  THE 'RATE' SUBROUTINE ACTUALLY COMPUTES THE DIRECTION FIELD V(I,1):
00600  REM  IT CALLS THE 'CONTROL' SUBROUTINE TO FIND THE CONTROL RATE PORTION.
00620  REM  RATE ALSO STORES THE PREVIOUS VALUE OF THE DIRECTION FIELD IN V(I,2)
00640  REM  AND COMPUTES THE DIFFERENCE V(I,3)=V(I,1)-V(I,2).  (OTHER NUMERICAL
00660  REM  METHODS MIGHT REQUIRE MORE BACK VALUES WHICH COULD BE STORED IN
00680  REM  OTHER COLUMNS OF V.)  'SOLVER' USES AN EULER STEP METHOD TO BEGIN,
00700  REM  THEN CALLS THE 'NEWTON' SUBROUTINE FOR NEWTON'S METHOD.THE MATRIX OF
00720  REM  PARTIAL DERIVATIVES IS FOUND BY THE 'FDERIV' SUBROUTINE.  WHEN THE
00740  REM  RMS RATE IS SMALL ENOUGH, 'SOLVER' STOPS THE ITERATION PROCESS, AND
00760  REM  THE MAIN PROGRAM SUMMARIZES RESULTS, THEN STEPS TO THE NEXT CONTROL
00780  REM  CASE.
00800  REM
00820  REM  DIMENSIONS OF VECTORS AND MATRICES:
00840  REM  IF THE NUMBER OF GROUPS G IS NOT 4, REPLACE 4 BY G IN DIM STATEMENTS
00860  REM  AND ALSO MODIFY PRINT LINES IN WHICH WE LABEL 4 GROUPS EXPLICITLY.
00880  DIM N(8),Y(8),D(8),K(8),P(8)
00900  DIM R(8),Z(8),W(8),U(8),E(8)
00920  DIM M(8,8),B(8,8),T(8,8),L(8,8)
```

```
00940   DIM A(8,8),V(8,3),F(8,1),X(8,1),H(8,1)
00960   DIM S(5,8),Q(11,8),O(1,8)
00980   REM
01000   REM - - - - - - - -    ENTER DATA    - - - - - - - - - - - - - - - - -
01020   REM
01040   REM   INDEXING:   WOMEN ODD, MEN EVEN
01060   REM   ORDERING:   GP1 WOMEN, GP2 MEN, GP3 WOMEN, GP4 MEN
01080   REM   G=NUMBER OF GROUPS
01100   G=8
01120   PRINT'INPUT G1=SELF INTERACTION FRACTION'
01140   INPUT G1
01200   REM  G2 SELECTS PRINTOUT: G2=0 FOR SUMMARY ONLY:  G2=1 FOR ITERATION INFO
01220   G2=0
01240   REM N(I)=RELATIVE SIZE OF GROUP I,IN ANY UNITS
01260   PRINT 'INPUT GROUP FRACTIONS N(I)'
01280   MAT INPUT N
01300   PRINT'INPUT ACTIVITY RATIOS OF GROUPS:WOMEN AND MEN'
01320   INPUT R8,R9
01340   REM  Y(I)=INITIAL GUESS OF PREVALENCE, THE INFECTIVE FRACTION OF GROUP I
01360   FOR I=1 TO G STEP 1
01380   READ Y(I)
01400   NEXT I
01420   DATA .5,.5,.1,.1,.1,.1,.1,.1
01440   REM  D(I)=DURATION OF DISEASE (AVERAGE) FOR GROUP I,IN ANY UNITS (MONTHS)
01460   PRINT 'INPUT DURATIONS D(I)'
01480   MAT INPUT D
01500   PRINT'INPUT PROB OF TRANS BY INF MAN/PROB OF TRANS BY INF WOMAN'
01520   INPUT R7
01540   REM  K(I)=NUMBER OF EFFECTIVE CONTACTS OF EACH GROUP I INFECTIVE
01560   REM   DURING THE INFECTIOUS PERIOD.
01580   PRINT'INPUT K(1)'
01600   INPUT K(1)
01620   K(2)=K(1)*D(2)/D(1)*R7*(N(1)+N(5)+(N(3)+N(7))/R8)/(N(2)+N(6)+(N(4)+N(8))/R9)
01640   K(3)=K(1)*D(3)/D(1)/R8
01650   K(4)=K(2)*D(4)/D(2)/R9
01660   K(5)=K(1)*D(5)/D(1)
01670   K(6)=K(2)*D(6)/D(2)
01680   K(7)=K(1)*D(7)/D(1)/R8
01690   K(8)=K(2)*D(8)/D(2)/R9
01780   REM - - - - - - - - - - - - - - - - -- - - - - - - - - - - - - - - - -
01800   REM PRELIMINARY CALCULATIONS
01820   MAT A=ZER(G,G)
01840   MAT F=ZER(G,1)
01860   MAT X=ZER(G,1)
01880   MAT H=ZER(G,1)
01900   MAT V=ZER(G,3)
01902   PRINT
01904   PRINT '####   RUN NUMBER        ####'
01918   PRINT
01920   PRINT ' ',' GONORRHEA CONTROL MODEL'
01940   PRINT ' ',' ----------------------- '
01980   REM
02000   REM    GO BUILD CONTACT MATRIX B
02020   REM
02040   GOSUB 8680
02060   REM
02080   REM **************** MAIN PROGRAM LOOP ****************************
02100   REM
02120   REM   R1 INDEXES THE CONTROL METHOD: R1=1 BEFORE THE CONTROLS ARE USED
02140   R1=1
```

```
02160   REM    GO TO THE SOLVER SUBROUTINE TO FIND THE SOLUTION
02180   GOSUB 4560
02200   IF G2=0 THEN 2280
02220   PRINT Y(1),Y(2),Y(3),Y(4),W5
02240   PRINT 'SOLUTION APPROACHED IS ON ABOVE LINE',' ','RMS RATE'
02280   IF R1>1 THEN 2680
02300   REM    SAVE VALUES FROM ORIGINAL CASE, BEFORE APPLYING CONTROLS
02320   S2=0
02340   S3=0
02360   S4=0
02380   S5=0
02400   FOR I=1 TO G STEP 1
02420   S(1,I)=Y(I)
02440   S(2,I)=Y(I)*N(I)
02460   S(3,I)=D(I)
02480   S(4,I)=S(2,I)/D(I)
02500   Z(I)=S(4,I)
02520   NEXT I
02540   FOR I=1 TO G-1 STEP 2
02560   S2=S2+Z(I+1)
02580   S3=S3+Z(I)
02600   S4=S4+S(2,I+1)
02620   S5=S5+S(2,I)
02640   NEXT I
02660   REM    STORE VALUES FOR PRINTING IN MATRIX Q
02680   Q2=0
02700   Q3=0
02720   Q4=0
02740   Q5=0
02760   FOR I=1 TO G STEP 1
02780   Q(1,I)=Y(I)
02800   Q(2,I)=(Q(1,I)-S(1,I))*100/S(1,I)
02820   Q(3,I)=Y(I)*N(I)
02840   Q(4,I)=(Q(3,I)-S(2,I))*100/S(2,I)
02860   Q(5,I)=D(I)
02880   Q(6,I)=1/(1/D(I)+C*R(I)/(Y(I)*N(I))+.1*C*P(I)/(Y(I)*N(I)))
02900   Q(7,I)=Q(3,I)/Q(6,I)
02920   Q(8,I)=(Q(7,I)-S(4,I))*100/S(4,I)
02940   Q(9,I)=C*R(I)+.1*C*P(I)
02960   Q(10,I)=N(I)
02980   Q(11,I)=K(I)
03000   NEXT I
03020   FOR I=1 TO G-1 STEP 2
03040   Q2=Q2+Q(7,I+1)
03060   Q3=Q3+Q(7,I)
03080   Q4=Q4+Q(3,I+1)
03100   Q5=Q5+Q(3,I)
03120   NEXT I
03160   ON R1 GOTO 3180,3220,3230,3260,3320
03180   PRINT '***********   RESULTS WITH NO EXTRA CURE RATE APPLIED   *********'
03200   GOTO 3340
03220   PRINT '**********  RESULTS FOR POPULATION SCREENING   ***************'
03221   PRINT 'THE NUMBER OF WOMEN DISCOVERED BY POPULATION SCREENING IS'
03222   PRINT 'SET EQUAL TO 10% OF THE INCIDENCE AND THE EFFECTIVE CONTACT'
03223   PRINT 'NUMBER IS CHOSEN SO THAT THIS POPULATION SCREENING CAUSES'
03224   PRINT 'APPROXIMATELY A 20% DECREASE IN INCIDENCE IN MEN.'
03225   GO TO 3340
03230   PRINT '*****RESULTS FOR SUPPLEMENTARY POPULATION SCREENING   *********'
03235   GO TO 3340
03260   PRINT '********   RESULTS FOR SUPPLEMENTARY INFECTEE TRACING   ***********'
```

```
03280    PRINT 'TRACE WOMEN REPORTED AS CONTACTEES OF DIAGNOSED MEN'
03300    GOTO 3340
03320    PRINT '********** RESULTS FOR SUPPLEMENTARY INFECTOR TRACING   **********'
03330    PRINT 'TRACE WOMEN REPORTED AS INFECTORS OF DIAGNOSED MEN'
03340    PRINT ' ','GP1 WOMEN','GP2 MEN','GP3 WOMEN','GP4 MEN'
03440    FOR I1=1 TO 11 STEP 1
03460    FOR I=1 TO G STEP 1
03480    O(1,I)=Q(I1,I)
03500    NEXT I
03540    ON I1 GOTO 3560,3600,3640,3680,3720,3760,3800,3840,3880,3920,3960
03560    PRINT 'PREVALENCE,FRAC':
03580    GOTO 3980
03600    PRINT 'CHANGE (%)     ':
03620    GOTO 3980
03640    PRINT 'PREVALENCE, POP':
03660    GOTO 3980
03680    PRINT 'CHANGE (%)     ':
03700    GOTO 3980
03720    PRINT 'DURATION,ORIG  ':
03740    GOTO 3980
03760    PRINT 'DURATION, NEW  ':
03780    GOTO 3980
03800    PRINT 'INCIDENCE,POP':
03820    GOTO 3980
03840    PRINT 'CHANGE (%)     ':
03860    GOTO 3980
03880    PRINT 'EXTRA CURE RATE':
03900    GOTO 3980
03920    IF R1>1 THEN 4040
03930    PRINT 'POPULAT. RATIOS':
03940    GOTO 3980
03960    PRINT 'CONTACT NUMBERS':
03980    PRINT O(1,1),O(1,2),O(1,3),O(1,4)
03990    PRINT' GPS 5,6,7,8',O(1,5),O(1,6),O(1,7),O(1,8)
04000    NEXT I1
04040    PRINT '   WOMEN,POP PREVALENCE =',Q5,'CHANGE (%)= ',(Q5-S5)*100/S5
04060    PRINT '   MEN,  POP PREVALENCE =',Q4,'CHANGE (%)=',(Q4-S4)*100/S4
04080    PRINT '   WOMEN,MONTHLY INCIDENCE =',Q3,'CHANGE (%)=',(Q3-S3)*100/S3
04100    PRINT '   MEN,  MONTHLY INCIDENCE =',Q2,'CHANGE (%)=',(Q2-S2)*100/S2
04105    IF R1=1 THEN 4200
04110    PRINT '   WOMEN,DISCOVERED BY POPULATION SCREENING=':C
04116    IF R1=2 THEN 4124
04120    PRINT '   WOMEN,DISCOVERED BY SUPPLEMENTARY CONTROL=':.1*C
04124    PRINT 'THE PROBABILITY THAT A WOMAN DISCOVERED BY POPULATION SCREENING'
04126    PRINT '     IS A GP1 MEMBER IS':R(1)
04130    IF R1>2 THEN 4136
04132    E2=Q2
04134    E3=Q3
04135    E7=Q5
04136    IF R1=2 THEN 4200
04138    PRINT 'THE PROBABILITY THAT A WOMAN DISCOVERED BY THE SUPPLEMENTARY'
04140    PRINT '     PROCEDURE IS A GP1 MEMBER IS':P(1)
04144    IF R1>3 THEN 4148
04146    C1=(E7-Q5)/E7*100
04148    C2=(E7-Q5)/E7*100
04150    PRINT 'IF THIS SUPPLEMENTARY CONTROL PROCEDURE INCREASES THE NUMBER OF'
04152    PRINT '     DISCOVERIES OF INFECTIOUS WOMEN BY 10%, THEN THE % DECREASE'
04154    PRINT '     IN PREVALENCE IN WOMEN IS':C2:'WHICH IS':C2/C1:'TIMES THE % '
04158    PRINT '  DECREASE IN PREVALENCE IN WOMEN DUE TO SUPPLEMENTARY POPULATIION'
04160    PRINT '     SCREENING.'
```

```
04170   E4=(E2-Q2)/(.1*C)
04172   E5=(E3-Q3)/(.1*C)
04174   PRINT 'THE DISCOVERY AND CURE OF ONE INFECTIOUS WOMAN BY THIS '
04176   PRINT '     SUPPLEMENTARY CONTROL PROCEDURE PREVENTS':E5:'CASES IN'
04178   PRINT '      WOMEN AND':E4:' CASES IN MEN.'
04200   PRINT '- - - - - - - - - - - - - - - - - - - - - - - - - - - - - -'
04210   IF R1<>1 THEN 4214
04212   PRINT
04214   IF R1<>3 THEN 4220
04215   PRINT
04220   R1=R1+1
04240   IF R1>5 THEN 4300
04280   GOTO 2180
04300   PRINT 'END OF PROGRAM'
04320   REM ********** END OF MAIN PROGRAM LOOP **************************
04340   STOP
04360   REM * * * * * * * * *  SUBROUTINES   * * * * * * * * * * * * * * *
04380   REM
04400   REM    ###############################################################
04410   REM
04420   REM                    SOLVER SUBROUTINE
04440   REM    THE PROCEDURE TO FIND PREVALENCES Y(I) BEGINS HERE
04460   REM
04480   REM  J2 COUNTS THE NUMBER OF ITERATIONS MADE SO FAR
04500   REM  J3 SELECTS THE METHOD:   J3=0 FOR EULER STEPS ALONG TRAJECTORY:
04520   REM                           J3=1 WHEN WE USE THE NEWTON METHOD
04540   REM  J5=0 IF WE DON'T WISH TO COMPUTE DIFFERENCES IN THE DIRECTION FIELD
04560   J2=0
04580   J3=0
04600   J5=0
04780   IF G2=0 THEN 5000
04800   PRINT 'PREVALENCES, AS FRACTIONS OF EACH GROUP'
04820   PRINT 'CORE WOMEN','CORE MEN','NONCORE WOMEN','NONCORE MEN','RMS RATE'
04840   PRINT Y(1),Y(2),Y(3),Y(4)
04860   GOTO 5000
04880   REM  THE ITERATION LOOP BEGINS HERE
04900   IF G2=0 THEN 5000
04920   PRINT Y(1),Y(2),Y(3),Y(4),W5
04940   REM
04960   REM  DECIDE WHICH SOLUTION METHOD TO APPLY NEXT
04980   REM
05000   IF J3=1 THEN 5400
05020   IF J4=G THEN 5030
05025   GO TO 5160
05030   IF J2>9 THEN 5060
05040   GOTO 5160
05060   J3=1
05080   IF G2=0 THEN 5160
05100   PRINT 'NEWTON METHOD BEGINS AFTER NEXT STEP'
05120   REM  THIS ITERATION PROCEDURE MAY BE USEFUL BEFORE NEWTON METHOD BEGINS:
05140   REM   GO TO THE RATE SUBROUTINE TO USE THE MODEL EQUATIONS
05160   GOSUB 6880
05180   J5=1
05200   FOR I=1 TO G STEP 1
05220   REM    Y(I)=(W(I)-S1*R(I))*D(I)/(1+D(I)*W(I))   IS AN ALTERNATIVE
05240   REM  EULER STEPS ALONG TRAJECTORY:
05260   Y(I)=Y(I)+.25*V(I,1)
05280   IF Y(I)>0 THEN 5320
05300   Y(I)=0
05320   NEXT I
```

```
05340    GOTO 5720
05360    REM WE GO TO THE NEWTON SUBROUTINE AFTER SAVING THE OLD SOLUTION IN E
05380    REM  AND THE OLD RMS RATE IN W7
05400    MAT E=Y
05420    W7=W5
05440    GOSUB 5820
05460    IF W5<W7 THEN 5720
05480    IF G2=0 THEN 5520
05500    PRINT 'USE ONLY HALF OF INCREMENT SINCE THE RMS RATE=',W5
05520    MAT Y=E+Y
05540    MAT Y=(.5)*Y
05560    GOSUB 5820
05580    IF W5<W7 THEN 5720
05600    MAT Y=E
05620    J3=0
05640    IF G2=0 THEN 5680
05660    PRINT 'USE STARTER AGAIN SINCE RMS RATE=',W5
05680    GOTO 5160
05700    REM   STEP UP ITERATION COUNTER
05720    J2=J2+1
05740    IF W5>1E-06 THEN 5750
05745    RETURN
05750    IF J2<300 THEN 4900
05755    PRINT 'WARNING: 300 ITERATIONS'
05760    RETURN
05780    REM     #############################################################
05800    REM
05820    REM       NEWTON SUBROUTINE STARTS HERE
05840    REM
05860    REM  WE WILL SOLVE F(X)=0, WHERE X CONTAINS PREVALENCES AND F CONTAINS
05880    REM   THEIR DIRECTION FIELD.
05900    REM
05920    FOR I=1 TO G STEP 1
05940    X(I,1)=Y(I)
05960    F(I,1)=V(I,1)
05980    NEXT I
06000    REM    GO TO SUB TO FORM PARTIAL DERIVATIVE MATRIX A
06020    GOSUB 6420
06040    REM
06060    REM    FIND NEW PREVALENCES X BY NEWTON METHOD, THEN STORE THEM IN Y
06080    REM
06100    MAT L=INV(A)
06120    MAT H=L*F
06140    MAT X=X-H
06160    REM
06180    FOR I=1 TO G STEP 1
06200    Y(I)=X(I,1)
06220    NEXT I
06240    GOSUB 6880
06260    RETURN
06280    REM    NEWTON ALGORITHM ENDS HERE
06300    REM
06320    REM     ###############################################################
06340    REM
06360    REM SUBROUTINE FDERIV:  FORM APPROXIMATE PARTIAL DERIVATIVE MATRIX A
06380    REM
06400    REM
06420    H9=.001
06440    J9=1
06460    Y(J9)=Y(J9)-H9
```

```
06480   GOSUB 6880
06500   J5=1
06520   Y(J9)=Y(J9)+2*H9
06540   GOSUB 6880
06560   REM  STORE PARTIALS:  MATRIX A APPROXIMATES DF
06580   FOR I=1 TO G STEP 1
06600   A(I,J9)=V(I,3)/(2*H9)
06620   NEXT I
06640   Y(J9)=Y(J9)-H9
06660   J9=J9+1
06680   J5=0
06700   IF J9<=G THEN 6460
06720   RETURN
06740   REM
06760   REM     ##############################################################
06780   REM
06800   REM     ****************** RATE SUBROUTINE  ***************************
06820   REM
06840   REM  WE COMPUTE THE DIRECTION FIELD V(I,1) USING THE MODEL EQUATIONS
06860   REM
06880   FOR I=1 TO G STEP 1
06920   W1=0
06940   FOR J1=1 TO G STEP 1
06960   W1=W1+B(I,J1)*Y(J1)
06980   NEXT J1
07000   W(I)=W1
07020   NEXT I
07040   REM
07060   REM GO TO SUB TO FORM EXTRA CURE RATES R(I)
07080   REM
07100   GOSUB 7580
07120   REM
07140   REM   COMPUTE V(I,1), TIME RATES OF CHANGE OF PREVALENCES  Y(I)
07160   REM
07180   W5=0
07200   J4=0
07220   FOR I=1 TO G STEP 1
07240   REM  ***********   MODEL EQUATIONS:   ********************
07260   V(I,1)=(1-Y(I))*W(I)-(Y(I)/D(I))-C*R(I)/N(I)-.1*C*P(I)/N(I)
07280   IF J5=0 THEN 7400
07300   REM CALCULATE CHANGES IN SLOPES
07320   V(I,3)=V(I,1)-V(I,2)
07340   IF J3=1 THEN 7400
07360   IF SGN(V(I,3))=SGN(V(I,1)) THEN 7400
07380   J4=J4+1
07400   V(I,2)=V(I,1)
07420   W5=W5+V(I,1)*V(I,1)
07440   NEXT I
07460   W5=SQR(W5)
07480   RETURN
07500   REM ***************************************************************
07520   REM  **************** CONTROL SUBROUTINE  ***********************
07540   REM  THESE LOOPS FORM EXTRA CURE RATES R(I) AND P(I)
07560   REM
07580   IF R1=1 THEN 7770
07586   IF R1>2 THEN 7720
07590   T1=C
07600   C=0
07620   FOR I=1 TO G-1 STEP 2
07640   Z(I)=Y(I)*N(I)*(1/D(I)+T1*R(I)/(Y(I)*N(I)))
```

```
07660    C=C+Z(I)
07680    NEXT I
07700    C=.1*C
07710    REM   C IS THE DISCOVERY RATE DUE TO POPULATION SCREENING
07720    IF R1<4 THEN 7880
07740    IF R1=4 THEN 8080
07760    IF R1=5 THEN 8360
07770    REM   UNCONTROLLED CASE
07780    FOR I=1 TO G STEP 1
07800    R(I)=0
07810    P(I)=0
07820    NEXT I
07860    RETURN
07880    REM POPULATION SAMPLING, WOMEN
07900    REM
07920    A3=0
07940    FOR I=1 TO G-1 STEP 2
07960    A3=A3+Y(I)*N(I)
07980    NEXT I
08000    FOR I=1 TO G-1 STEP 2
08020    R(I)=Y(I)*N(I)/A3
08040    NEXT I
08050    IF R1=2 THEN 8060
08052    FOR I=1 TO G-1 STEP 2
08054    P(I)=R(I)
08056    NEXT I
08060    RETURN
08080    REM INFECTEE SAMPLING, WOMEN
08100    A5=0
08120    FOR I=1 TO G-1 STEP 2
08140    A4=0
08160    FOR I9=2 TO G STEP 2
08180    A4=A4+M(I,I9)*(1-Y(I))*Y(I9)*N(I9)*K(I9)/D(I9)
08200    NEXT I9
08220    P(I)=A4
08240    A5=A5+P(I)
08260    NEXT I
08280    FOR I=1 TO G-1 STEP 2
08300    P(I)=P(I)/A5
08320    NEXT I
08340    RETURN
08360    REM INFECTOR SAMPLING, WOMEN
08380    A5=0
08400    FOR I=1 TO G-1 STEP 2
08420    A4=0
08440    FOR I9=2 TO G STEP 2
08460    A4=A4+M(I9,I)*(1-Y(I9))
08480    NEXT I9
08500    P(I)=A4*N(I)*Y(I)*K(I)/D(I)
08520    A5=A5+P(I)
08540    NEXT I
08560    FOR I=1 TO G-1 STEP 2
08580    P(I)=P(I)/A5
08600    NEXT I
08620    RETURN
08640    REM    ######################################################################
08660    REM
08680    MAT M=ZER(G,G)
08700    MAT T=ZER(G,G)
08720    MAT B=ZER(G,G)
```

```
08760   REM SET UP PROPORTIONATE MIXING MODEL
08780   A1=0
08800   A2=0
08820   FOR I=1 TO G STEP 1
08840   U(I)=N(I)*K(I)/D(I)
08860   NEXT I
08880   FOR I=2 TO G STEP 2
08900   A1=A1+U(I-1)
08920   A2=A2+U(I)
08940   NEXT I
08960   FOR I9=1 TO G-1 STEP 2
08980   FOR I=2 TO G STEP 2
09000   M(I-1,I9+1)=U(I-1)/A1
09020   M(I,I9)=U(I)/A2
09040   NEXT I
09060   NEXT I9
09080   IF G1=0 THEN 9440
09120   MAT L=ZER(G,G)
09130   FOR K8=3 TO 7 STEP 4
09140   FOR K9=1 TO 2
09150   U8=U(K9)+U(K9+4)
09160   L(K9,K8-K9)=U(K9)/U8
09170   L(K9+4,K8-K9)=U(K9+4)/U8
09180   U9=U(K9+2)+U(K9+6)
09190   L(K9+2,K8-K9+2)=U(K9+2)/U9
09200   L(K9+6,K8-K9+2)=U(K9+6)/U9
09210   NEXT K9
09220   NEXT K8
09250   MAT L=(G1)*L
09260   MAT M=(1-G1)*M
09270   MAT M=M+L
09440   FOR I9=1 TO G STEP 1
09460   FOR I=1 TO G STEP 1
09480   T(I,I9)=M(I,I9)*K(I9)
09500   B(I,I9)=T(I,I9)*N(I9)/(D(I9)*N(I))
09520   NEXT I
09540   NEXT I9
09745   IF G1>0 THEN 9780
09748   PRINT
09750   PRINT 'PROB. OF TRANS. BY INF. MAN/PROB. OF TRANS. BY INF. WOMAN=':R7
09752   PRINT 'SEXUAL ACTIVITY OF GP1 WOMAN/SEXUAL ACT. OF GP3 WOMAN=':R8
09754   PRINT 'SEXUAL ACTIVITY OF GP2 MAN/SEXUAL ACTIVITY OF GP4 MAN=':R9
09756   P1=0
09758   P2=0
09760   FOR I=2 TO G STEP 2
09762   P1=P1+M(I,1)*K(I)
09764   P2=P2+M(I-1,2)*K(I-1)
09766   NEXT I
09768   K4=P1*P2
09770   PRINT'EFFECTIVE CONTACT NUMBER=':K4
09772   PRINT
09780   IF G2=0 THEN 9880
09800   PRINT 'MIXING MATRIX: M(I,J)=FRAC OF EFF. CONTACTS OF J THAT ARE WITH I'
09840   PRINT 'CONTACT MATRIX B'
09880   RETURN
09900   END
```

SAMPLE RUN OF GCCONT SHOWING
INPUT PARAMETER VALUES AND OUTPUT TABLES

```
OK, BASIC GCCONT.8
BASIC REV19.0
INPUT G1=SELF INTERACTION FRACTION
!.2
INPUT GROUP FRACTIONS N(I)
!.012,.002,.588,.098,.008,.018,.392,.882
INPUT ACTIVITY RATIOS OF GROUPS:WOMEN AND MEN
!10,10
INPUT DURATIONS D(I)
!6,6,6,6,.5,.5,.5,.5
INPUT PROB OF TRANS BY INF MAN/PROB OF TRANS BY INF WOMAN
!2
INPUT K(1)
!9.238
```

RUN NUMBER

GONORRHEA CONTROL MODEL

********** RESULTS WITH NO EXTRA CURE RATE APPLIED *********

	GP1 WOMEN	GP2 MEN	GP3 WOMEN	GP4 MEN
PREVALENCE,FRAC	.497361	.548178	.0496891	.0622565
GPS 5,6,7,8	.0761769	.0918213	.00433837	.00550203
CHANGE (%)	0	0	0	0
GPS 5,6,7,8	0	0	0	0
PREVALENCE, POP	.00596833	.00109636	.0292172	.00610114
GPS 5,6,7,8	.000609415	.00165278	.00170064	.00485279
CHANGE (%)	0	0	0	0
GPS 5,6,7,8	0	0	0	0
DURATION,ORIG	6	6	6	6
GPS 5,6,7,8	.5	.5	.5	.5
DURATION, NEW	6	6	6	6
GPS 5,6,7,8	.5	.5	.5	.5
INCIDENCE,POP	.000994722	.000182726	.00486953	.00101686
GPS 5,6,7,8	.00121883	.00330557	.00340128	.00970558
CHANGE (%)	0	0	0	0
GPS 5,6,7,8	0	0	0	0
EXTRA CURE RATE	0	0	0	0
GPS 5,6,7,8	0	0	0	0
POPULAT. RATIOS	.012	.002	.588	.098
GPS 5,6,7,8	.008	.018	.392	.882

```
CONTACT NUMBERS 9.238          18.476          .9238          1.8476
  GPS 5,6,7,8  .769833         1.53967         .0769833       .153967
   WOMEN,POP PREVALENCE =      .0374956     CHANGE (%)=  0
   MEN,   POP PREVALENCE =     .0137031     CHANGE (%)=  0
   WOMEN,MONTHLY INCIDENCE =  .0104844      CHANGE (%)=  0
   MEN,   MONTHLY INCIDENCE = .0142107      CHANGE (%)=  0
- - - - - - - - - - - - - - - - - - - - - - - - - - - - - - - - - -
```

********** RESULTS FOR POPULATION SCREENING ***************
THE NUMBER OF WOMEN DISCOVERED BY POPULATION SCREENING IS
SET EQUAL TO 10% OF THE INCIDENCE AND THE EFFECTIVE CONTACT
NUMBER IS CHOSEN SO THAT THIS POPULATION SCREENING CAUSES
APPROXIMATELY A 20% DECREASE IN INCIDENCE IN MEN.

	GP1 WOMEN	GP2 MEN	GP3 WOMEN	GP4 MEN
PREVALENCE,FRAC	.412828	.496635	.0352891	.0497078
GPS 5,6,7,8	.0644987	.0759727	.00357429	.00434007
CHANGE (%)	-16.9963	-9.4026	-28.9804	-20.1565
GPS 5,6,7,8	-15.3304	-17.2603	-17.6121	-21.1188
PREVALENCE, POP	.00495394	.000993269	.02075	.00487136
GPS 5,6,7,8	.000515989	.00136751	.00140112	.00382794
CHANGE (%)	-16.9963	-9.40262	-28.9804	-20.1565
GPS 5,6,7,8	-15.3304	-17.2603	-17.6121	-21.1188
DURATION,ORIG	6	6	6	6
GPS 5,6,7,8	.5	.5	.5	.5
DURATION, NEW	5.01689	6	5.01689	6
GPS 5,6,7,8	.491966	.5	.491966	.5
INCIDENCE,POP	.000987452	.000165545	.00413602	.000811894
GPS 5,6,7,8	.00104883	.00273502	.002848	.00765588
CHANGE (%)	-.730848	-9.40261	-15.0633	-20.1565
GPS 5,6,7,8	-13.9478	-17.2603	-16.2667	-21.1188
EXTRA CURE RATE	.000161796	0	.000677695	0
GPS 5,6,7,8	1.68522E-05	0	4.57607E-05	0

```
   WOMEN,POP PREVALENCE =      .027621      CHANGE  (%)=  -26.3353
   MEN,   POP PREVALENCE =     .0110601     CHANGE  (%)=  -19.2876
   WOMEN,MONTHLY INCIDENCE = .00902031      CHANGE  (%)=  -13.9642
   MEN,   MONTHLY INCIDENCE = .0113683      CHANGE  (%)=  -20.0018
   WOMEN,DISCOVERED BY POPULATION SCREENING= .000902103
```
THE PROBABILITY THAT A WOMAN DISCOVERED BY POPULATION SCREENING
    IS A GP1 MEMBER IS .179354
- -
*****RESULTS FOR SUPPLEMENTARY POPULATION SCREENING *********

	GP1 WOMEN	GP2 MEN	GP3 WOMEN	GP4 MEN
PREVALENCE,FRAC	.400892	.488705	.0335746	.0480485
GPS 5,6,7,8	.0628551	.0737751	.00347012	.00418852
CHANGE (%)	-19.3963	-10.8492	-32.4308	-22.8217
GPS 5,6,7,8	-17.488	-19.6536	-20.0132	-23.8732
PREVALENCE, POP	.0048107	.000977409	.0197419	.00470875
GPS 5,6,7,8	.00050284	.00132795	.00136029	.00369428
CHANGE (%)	-19.3963	-10.8492	-32.4308	-22.8217
GPS 5,6,7,8	-17.4881	-19.6536	-20.0132	-23.8732
DURATION,ORIG	6	6	6	6
GPS 5,6,7,8	.5	.5	.5	.5
DURATION, NEW	4.89639	6	4.89639	6
GPS 5,6,7,8	.490782	.5	.490782	.5
INCIDENCE,POP	.000982498	.000162902	.00403192	.000784792
GPS 5,6,7,8	.00102457	.00265591	.00277168	.00738855
CHANGE (%)	-1.22889	-10.8492	-17.2012	-22.8217
GPS 5,6,7,8	-15.9383	-19.6536	-18.5108	-23.8732
EXTRA CURE RATE	.000180716	0	.000741609	0
GPS 5,6,7,8	1.88894E-05	0	5.10997E-05	0

```
WOMEN,POP PREVALENCE =     .0264157     CHANGE (%)=   -29.5499
MEN,  POP PREVALENCE =     .0107084     CHANGE (%)=   -21.8541
WOMEN,MONTHLY INCIDENCE = .00881066     CHANGE (%)=   -15.9639
MEN,  MONTHLY INCIDENCE = .0109922      CHANGE (%)=   -22.649
WOMEN,DISCOVERED BY POPULATION SCREENING= .000902103
WOMEN,DISCOVERED BY SUPPLEMENTARY CONTROL= 9.02103E-05
```
THE PROBABILITY THAT A WOMAN DISCOVERED BY POPULATION SCREENING
    IS A GP1 MEMBER IS .182115
THE PROBABILITY THAT A WOMAN DISCOVERED BY THE SUPPLEMENTARY
    PROCEDURE IS A GP1 MEMBER IS .182115
IF THIS SUPPLEMENTARY CONTROL PROCEDURE INCREASES THE NUMBER OF
    DISCOVERIES OF INFECTIOUS WOMEN BY 10%, THEN THE % DECREASE
    IN PREVALENCE IN WOMEN IS 4.36384 WHICH IS 1 TIMES THE %
DECREASE IN PREVALENCE IN WOMEN DUE TO SUPPLEMENTARY POPULATIION
    SCREENING.
THE DISCOVERY AND CURE OF ONE INFECTIOUS WOMAN BY THIS
    SUPPLEMENTARY CONTROL PROCEDURE PREVENTS 2.32404 CASES IN
    WOMEN AND 4.17013 CASES IN MEN.

- -

******** RESULTS FOR SUPPLEMENTARY INFECTEE TRACING ***********
TRACE WOMEN REPORTED AS CONTACTEES OF DIAGNOSED MEN

	GP1 WOMEN	GP2 MEN	GP3 WOMEN	GP4 MEN
PREVALENCE,FRAC	.404722	.491104	.0341616	.0485614
GPS 5,6,7,8	.0628028	.0744338	.00346952	.00423532
CHANGE (%)	-18.6262	-10.4116	-31.2493	-21.9978
GPS 5,6,7,8	-17.5567	-18.9363	-20.027	-23.0227
PREVALENCE, POP	.00485666	.000982207	.020087	.00475902
GPS 5,6,7,8	.000502422	.00133981	.00136005	.00373555
CHANGE (%)	-18.6262	-10.4116	-31.2493	-21.9978
GPS 5,6,7,8	-17.5567	-18.9363	-20.027	-23.0227
DURATION,ORIG	6	6	6	6
GPS 5,6,7,8	.5	.5	.5	.5
DURATION, NEW	4.93699	6	4.94349	6
GPS 5,6,7,8	.486602	.5	.486607	.5
INCIDENCE,POP	.000983729	.000163701	.00406333	.00079317
GPS 5,6,7,8	.00103251	.00267962	.00279497	.0074711
CHANGE (%)	-1.1051	-10.4116	-16.556	-21.9978
GPS 5,6,7,8	-15.2867	-18.9363	-17.8259	-23.0227
EXTRA CURE RATE	.000174287	0	.000715494	0
GPS 5,6,7,8	2.76677E-05	0	7.48653E-05	0

```
  WOMEN,POP PREVALENCE =     .0268062     CHANGE (%)=   -28.5085
  MEN,  POP PREVALENCE =     .0108166     CHANGE (%)=   -21.0645
  WOMEN,MONTHLY INCIDENCE = .00887454     CHANGE (%)=   -15.3545
  MEN,  MONTHLY INCIDENCE = .0111076      CHANGE (%)=   -21.8366
  WOMEN,DISCOVERED BY POPULATION SCREENING= .000902103
  WOMEN,DISCOVERED BY SUPPLEMENTARY CONTROL= 9.02103E-05
```
THE PROBABILITY THAT A WOMAN DISCOVERED BY POPULATION SCREENING
    IS A GP1 MEMBER IS .182115
THE PROBABILITY THAT A WOMAN DISCOVERED BY THE SUPPLEMENTARY
    PROCEDURE IS A GP1 MEMBER IS .110849
IF THIS SUPPLEMENTARY CONTROL PROCEDURE INCREASES THE NUMBER OF
    DISCOVERIES OF INFECTIOUS WOMEN BY 10%, THEN THE % DECREASE
    IN PREVALENCE IN WOMEN IS 2.95006 WHICH IS .676024 TIMES THE %
DECREASE IN PREVALENCE IN WOMEN DUE TO SUPPLEMENTARY POPULATIION
    SCREENING.
THE DISCOVERY AND CURE OF ONE INFECTIOUS WOMAN BY THIS
    SUPPLEMENTARY CONTROL PROCEDURE PREVENTS 1.61584 CASES IN
    WOMEN AND 2.89047 CASES IN MEN.

- -

```
********* RESULTS FOR SUPPLEMENTARY INFECTOR TRACING  *********
TRACE WOMEN REPORTED AS INFECTORS OF DIAGNOSED MEN
                 GP1 WOMEN      GP2 MEN        GP3 WOMEN      GP4 MEN
PREVALENCE,FRAC .377769         .474841        .0322703       .045732
  GPS 5,6,7,8  .0600718         .0700691       .00332028      .00397775
CHANGE (%)       -24.0453       -13.3783       -35.0557       -26.5426
  GPS 5,6,7,8  -21.1417         -23.6898       -23.4671       -27.704
PREVALENCE, POP .00453323       .000949681     .0189749       .00448174
  GPS 5,6,7,8  .000480574       .00126124      .00130155      .00350838
CHANGE (%)       -24.0453       -13.3783       -35.0557       -26.5426
  GPS 5,6,7,8  -21.1417         -23.6898       -23.4671       -27.704
DURATION,ORIG   6               6              6              6
  GPS 5,6,7,8  .5               .5             .5             .5
DURATION, NEW   4.63736         6              4.91406        6
  GPS 5,6,7,8  .488171          .5             .49092         .5
INCIDENCE,POP .000977546        .00015828      .00386135      .000746956
  GPS 5,6,7,8  .000984439       .00252249      .00265124      .00701675
CHANGE (%)       -1.72675       -13.3783       -20.7038       -26.5426
  GPS 5,6,7,8  -19.2308         -23.6898       -22.0516       -27.704
EXTRA CURE RATE .000222008      0              .000698867     0
  GPS 5,6,7,8  2.32913E-05      0              4.8147E-05     0
    WOMEN,POP PREVALENCE =      .0252903       CHANGE (%)=    -32.5514
    MEN,  POP PREVALENCE =      .010201        CHANGE (%)=    -25.5566
    WOMEN,MONTHLY INCIDENCE = .00847458        CHANGE (%)=    -19.1693
    MEN,  MONTHLY INCIDENCE = .0104445         CHANGE (%)=    -26.5029
    WOMEN,DISCOVERED BY POPULATION SCREENING= .000902103
    WOMEN,DISCOVERED BY SUPPLEMENTARY CONTROL= 9.02103E-05
THE PROBABILITY THAT A WOMAN DISCOVERED BY POPULATION SCREENING
     IS A GP1 MEMBER IS .182115
THE PROBABILITY THAT A WOMAN DISCOVERED BY THE SUPPLEMENTARY
     PROCEDURE IS A GP1 MEMBER IS .639851
IF THIS SUPPLEMENTARY CONTROL PROCEDURE INCREASES THE NUMBER OF
     DISCOVERIES OF INFECTIOUS WOMEN BY 10%, THEN THE % DECREASE
     IN PREVALENCE IN WOMEN IS 8.43827 WHICH IS 1.93368 TIMES THE %
   DECREASE IN PREVALENCE IN WOMEN DUE TO SUPPLEMENTARY POPULATIION
     SCREENING.
THE DISCOVERY AND CURE OF ONE INFECTIOUS WOMAN BY THIS
     SUPPLEMENTARY CONTROL PROCEDURE PREVENTS 6.04948 CASES IN
     WOMEN AND 10.2412  CASES IN MEN.
- - - - - - - - - - - - - - - - - - - - - - - - - - - - - - - - -
END OF PROGRAM

STOPPED AT LINE 4340
```

REFERENCES

American Social Health Association, 1975. Todays VD Control Problem.

Anderson, R.M. and R.M. May, 1979. Population biology of infectious diseases I, Nature 280, 361-367.

Anderson, R.M. and R.M. May, 1982. Directly transmitted infectious diseases: control by vaccination, Science 215, 1053-1060.

Aronsson, G. and I. Mellander, 1980. A deterministic model in biomathematics: asymptotic behavior and threshold conditions, Math. Biosci. 49, 207-222.

Bailey, N.T.J., 1975. The Mathematical Theory of Infectious Diseases, 2nd ed., Hafner Press, New York.

Bailey, N.T.J., 1979. Introduction to the modeling of venereal disease, J. Math. Biology 8, 301-322.

Blount, J.H., 1979. Private communication.

Braun, M., 1975. Differential Equations and Their Applications, Springer-Verlag, New York.

Bureau of the Census, 1977. U.S. Department of Commerce. Current population reports: Population estimates and projections. Projections of the population of the United States: 1977-2050. Series P25, No. 704.

Centers for Disease Control, 1979a. Results of culture testing for gonorrhea - United States 1978, Morbidity and Mortality Weekly Report 28, 290-291.

Centers for Disease Control, 1979b. Gonorrhea - United States, Morbidity and Mortality Weekly Report 28, 533-534.

Centers for Disease Control, 1980a. Penicillinase-producing Neisseria gonorrhoeae - Los Angeles, California, Morbidity and Mortality Weekly Report 29, 541-543.

Centers for Disease Control, 1980b. Pelvic inflammatory disease - United States, Morbidity and Mortality Weekly Report 28, 605-607.

Centers for Disease Control, 1981a. Annual Summary 1980: reported morbidity and mortality in the United States, Morbidity and Mortality Weekly Report 29 (54).

Centers for Disease Control, 1981b. Spectinomycin-resistant Penicillinase-producing Neisseria gonorhoeae - Cal., Morbidity and Mortality Weekly Report 30, 221-222.

Centers for Disease Control, 1982. Global distribution of Penicillinase-producing Neisseria gonorrhoeae (PPNG), Morbidity and Mortality Weekly Report 3, 1-3.

Centers for Disease Control, 1983. Sexually Transmitted Disease Statistical Letter 1982.

Conrad, G.L., G.S. Klevis, B. Rush, W.W. Darrow, 1981. Sexually transmitted diseases among prostitutes and other sexual offenders. Sex. Trans. Dis. 8, 241-244.

Constable, G.M., 1975. The problem of V.D. modelling, Bull. Inst. Int. Statist. 106-2, 256-263.

Cooke, K.L., 1976. An epidemic equation with immigration, Math. Biosci. 29, 135-138.

Cooke, K.L., 1982. Models for endemic infections with asymptomatic cases. I One group, Math. Modelling 3, 1-15.

Cooke, K.L., 1984. Infection models with asymptomatics, in proceedings of the second IMACS International Symposium on Biomedical Systems Modeling, Bethesda, Maryland, August 1984, C. DeLisi and J. Eisenfeld, eds., North Holland, Amsterdam.

Cooke, K.L. and J.A. Yorke, 1973. Some equations modelling growth processes and gonorrhea epidemics, Math. Biosci. 16, 75-101.

Cornelius III, C.E., 1971. Seasonality of gonorrhea in the United States. HSMHA Health Rep. 86, 157-160.

Curran, J.W., 1980. Economic consequences of pelvic inflammatory disease in the United States. Am. J. Obstet. Gynecol. 138, 848-851.

Darrow, W.W., D. Barrett, K. Jay et al., 1981. The gay report on sexually transmitted diseases, Am. J. Publ. Health 71, 1004-1011.

Darrow, W.W., and M.L. Pauli, 1983. Prostitution and STD, in Sexually Transmitted Diseases, Weisner, Mardh, Holmes, Sparling, eds., McGraw-Hill, New York.

Dietz, K., 1975. Transmission and control of arbovirus diseases, in Epidemiology, SIMS 1974 Utah Conference Proceedings, SIAM, Philadelphia, 104-121.

Dietz, K., 1976. The incidence of infectious diseases under the influence of seasonal fluctuations, in Mathematical Models in Medicine, Lecture Notes in Biomathematics, No. 11, Springer-Verlag, New York, 1-15.

Henderson, R.H., 1975a. National strategies to control gonorrhea, Minutes of the Venereal Disease Control Advisory committee meeting of September 18-19.

Henderson, R.H., 1975b. Dear Colleagues letter of October 1 from Director of Venereal Disease Control Division, Centers for Disease Control, Attachment #2, Commentary on national strategies to control gonorrhea.

Hethcote, H.W., 1973. Asymptotic behavior in a deterministic epidemic model, Bull. Math. Biology 35, 607-614.

Hethcote, H.W., 1974. Asymptotic behavior and stability in epidemic models, in Mathematical Problems in Biology, P. van den Driessche, ed., Lecture Notes in Biomathematics, No. 2, Springer-Verlag, New York, 83-92.

Hethcote, H.W., 1975. Mathematical models for the spread of infectious diseases, in Epidemiology, SIMS 1974 Utah Conference Proceedings, D. Ludwig and K.L. Cooke, eds., SIAM, Philadelphia, 122-131.

Hethcote, H.W., 1976. Qualitative analyses for communicable disease models, Math. Biosci., 28, 335-356.

Hethcote, H.W., 1978. An immunization model for a heterogeneous population, Theor. Pop. Biol. 14, 338-349.

Hethcote, H.W., 1983. Measles and rubella in the United States, Am. J. Epid. 117, 2-13.

Hethcote, H.W., H.W. Stech and P. van den Driessche, 1981a. Nonlinear oscillations in epidemic models, SIAM J. Applied Math. 40, 1-9.

Hethcote, H.W., H.W. Stech and P. van den Driessche, 1981b. Stability analysis for models of diseases without immunity, J. Math. Biol. 13, 185-198.

Hethcote, H.W., H.W. Stech and P. van den Driessche, 1981c. Periodicity and stability in epidemic models: a survey, in Differential Equations and Applications in Ecology, Epidemics and Population Problems. S. Busenberg and K.L. Cooke, eds. Academic Press, New York, 65-82.

Hethcote, H.W. and D.W. Tudor, 1980. Integral equation models for endemic infectious diseases, J. Math. Biol. 9, 37-47.

Hethcote, H.W. and P. Waltman, 1973. Optimal vaccination schedules in a deterministic epidemic model, Math. Biosci. 18, 365-382.

Hethcote, H.W., J.A. Yorke and A. Nold, 1982. Gonorrhea modeling: a comparison of control methods, Math. Biosci. 58, 93-109.

Hirsch, M. W., 1984. The differential equations approach to dynamical systems. Bull. Amer. Math. Soc., 11, 1-64.

Jones, O.G., 1976. Private communication.

Judson, F.N., K.A. Penley, M.E. Robinson et al., 1980. Comparative prevalence rates of sexually transmitted diseases in heterosexual and homosexual men. Am. J.. Epidemiol. 112, 836-843.

Kemper, J.T., 1978. The effects of asymptomatic attacks on the spread of infectious diseases: A deterministic model, Bull. Math. Biology 40, 707-718.

Kemper, J.T., 1980. On the identification of superspreaders for infectious diseases. Math. Biosci. 48, 111-127.

Kramer, M.A. and G.H. Reynolds, 1981. Evaluation of a gonorrhea vaccine and other gonorrhea control strategies based on computer simulation modeling, in Differential Equations and Applications in Ecology, Epidemics and Population Problems. S. Busenberg and K.L. Cooke, eds. Academic Press, New York, 97-114.

Lajmanovich, A. and J.A. Yorke, 1976. A deterministic model for gonorrhea in a nonhomogeneous population, Math. Biosci. 28, 221-236.

London, W.P. and J.A. Yorke, 1973. Recurrent outbreaks of measles, chickenpox and mumps: I Seasonal variation in contact rates. Am. J. Epidemiol, 98, 453-468.

Mahoney, J.F. et al., 1942. Culture studies in chronic gonorrhea of women, Amer. J. Symph., Gon., and V.D. 26, 38-47.

Marx, J.L., 1980. Vaccinating with bacterial pili, Science 209, 1103-1106.

May, R.M., 1981. The transmission and control of gonorrhea, Nature 291, 376-377.

Miles, J.R., 1978. Gonorrhea control: rescreening, Attachment #4 in Dear Colleagues letter of October 18 from P.J. Wiesner, Director of Veneral Disease Control Division, Centers for Disease Control.

Miller, R.K. and A.N. Michel, 1982. Ordinary Differential Equations, Academic Press, New York.

Nallaswamy, R. and J.B. Shukla, 1982. Effects of dispersal on the stability of a gonorrhea endemic model. Math. Biosci. 61, 63-72.

National Institute of Allergy and Infectious Diseases, 1980. Sexually transmitted diseases: 1980 status report, NIH Publication No. 81-2213.

Nold. A., 1979. The infectee number for communicable diseases, Math. Biosci. 46, 131-139.

Nold, A., 1980. Heterogeneity in disease transmission modeling. Math. Biosci. 52, 227-240.

Phillips, L., J.J. Potterat, R.B. Rothenberg et al., 1980. Focused interviewing in gonorrhea control, Amer. J. Publ. Hlth. 70, 705-708.

Potterat, J.J., R.B. Rothenberg and D.C. Bross, 1979. Gonorrhea in street prostitutes: Epidemiologic and legal implications. Sex. Trans. Dis. 6, 58-63.

Potterat, J.J., L. Phillips, R.B. Rothenberg, W.W. Darrow, 1980. Gonococcal pelvic inflammatory disease: case-finding observations, Am. J. Obstet. Gyn. 137, 1101-1104.

Potterat, J.J., R.B. Rothenberg, D. Woodhouse et at., 1983. Gonorrhea as a social disease. Presented at the Fifth Meeting of the International Society for STD Research, Seattle, Washington, August 1-3, 1983.

Rein, M.F., 1977. Epidemiology of gonococcal infections, in Gonococcus, R.B. Roberts, ed., Wiley, New York, 1-47.

Rendtorff, R.C. et al., 1974. Economic consequences of gonorrhea in women, J. Amer. VD Assoc. 1, 40-47.

Reynolds, G.H. and Y.K. Chan, 1974. A control model for gonorrhea, Bull. Inst. Int. Statist. 106-2, 264-279.

Rothenberg, R.B., 1982. High risk groups: Identification and intervention. Presented at the WHO/PAHO Scientific Working Group on the Control of Sexually Transmitted Diseases, Washington, D.C., April 26-30, 1982.

Rothenberg, R.B., 1983. The geography of gonorrhea: Empirical demonstration of core group transmission. Am. J. Epidemiol. 117, 688–694.

St. John, R.K. and J.W. Curran, 1978. Epidemiology of gonorrhea, Sexually Transmitted Diseases 5, 81–82.

Scientific American, 1976. Gonorrhea Resurgent, Vol. 234, No. 6, June, p. 50.

Shearer, L., 1983. VD Vaccine, Parade Magazine of April 24, 10.

Strauss, A. and J. A. Yorke, 1967. On asymptotically autonomous differential equations, Math. Systems Th. 1, 175–182.

Strauss, A. and J. A. Yorke, 1969. Perturbing uniform asymptotically stable nonlinear systems, J. Diff. Equ. 6, 452–483.

Thieme, H. R., 1982. Global asymptotic stability in epidemic models, Equadiff 1982.

Wichmann, H.E., 1979. Asymptotic behavior and stability in four models of venereal disease, J. Math. Biology 8, 301–322.

Wiesner, P.J., 1979. Dear Colleagues letter of April 27 from Director of Venereal Disease Control Division, Centers for Disease Control, Attachment #7, Identifying high risk patients by census tract.

Wiesner, P.J., 1980a. Dear Colleagues letter of March 10 from Director of Venereal Disease Control Division, Centers for Disease Control.

Wiesner, P.J., 1980b. Dear Colleagues letter of December 15 from Director of Venereal Disease Control Division, Centers for Disease Control.

Wiesner, P.J. and S.E. Thompson III, 1980. Gonococcal diseases, Disease-a-Month 27, No. 5, 1–44.

World Health Organization, 1978. Neisseria Gonorrhoeae and Gonococcal Infections, Technical Report Series 616, Geneva.

Yorke, J.A., H.W. Hethcote, and A. Nold, 1978. Dynamics and control of the transmission of gonorrhea, Sexually Transmitted Diseases 5, 51–56.

Yorke, J.A. and W.P. London, 1973. Recurrent outbreaks of measles, chickenpox and mumps: II Systematic differences in contact rates and stochastic effects. Am. J. Epidemiol. 98, 469–482.

Yorke, J.A., N. Nathanson, G. Pianigiani, and J. Martin, 1979. Seasonality and the requirements for perpetuation and eradication of viruses in populations, Amer. J. Epidemiol. 109, 103–123.

Zaidi, A.A., S.O. Aral, G.H. Reynolds et. al., 1983. Gonorrhea in the U.S.: 1967–1979. Sex. Trans. Dis. 10, 72–76.

Biomathematics

Managing Editor: **S.A.Levin**

Volume 9
W.J.Ewens

Mathematical Population Genetics

1979. 4 figures, 17 tables. XII, 325 pages.
ISBN 3-540-09577-2

This graduate level monograph considers the mathematical theory of population genetics, emphasizing aspects relevant to evolutionary studies. It contains a definitive and comprehensive discussion of relevant areas with references to the essential literature. The sound presentation and excellent exposition make this book a standard for population geneticists interested in the mathematical foundations of their subject as well as for mathematicians involved with genetic ecolutionary processes.

Volume 10
A.Okubo

Diffusion and Ecological Problems: Mathematical Models

1980. 114 figures, 6 tables. XIII, 254 pages.
ISBN 3-540-09620-5

This is the first comprehensive book on mathematical models of diffusion in an ecological context. Directed towards applied mathematicians, physicists and biologists, it gives a sound, biologically oriented treatment of the mathematics and physics of diffusion.

Volume 11
B.G.Mirkin, S.N.Rodin

Graphs and Genes

Translated from the Russian by H.L.Beus
1984. 46 figures. XIV, 197 pages. ISBN 3-540-12657-0

Contents: Graphs in the analysis of gene structure. – Graphs in the analysis of gene semantics. – Graphs in the analysis of gene evolution. – Epilogue: Cryptographic problems in genetics. – Appendix: Some notions about graphs. – References. – Index of genetics terms. – Index of mathematical terms.

Springer-Verlag
Berlin
Heidelberg
NewYork
Tokyo

Journal of Mathematical Biology

ISSN 0303-6812   Title No. 285

Editorial Board:

H. T. Banks, Providence, RI; H. J. Bremermann, Berkeley, CA; J. D. Cowan, Chicago, IL; J. Gani, Lexington, KY; K. P. Hadeler (Managing Editor), Tübingen; F. C. Hoppensteadt, Salt Lake City, UT; S. A. Levin (Managing Editor), Ithaca, NY; D. Ludwig, Vancouver; L. J. D. Murray, Oxford; L. T. Nagylaki, Chicago, IL; L. A. Segel, Rehovot; D. Varjú, Tübingen in cooperation with a distinguished advisory board.

For mathematicians and biologists working in a wide spectrum of fields, the **Journal of Mathematical Biology** publishes:

- papers in which mathematics in used to better understand biological phenomena
- mathematical papers inspired by biological research and
- papers which yield new experimental data bearing on mathematical models.

Contributions also discuss related areas of medicine, chemistry, and physics.

Articles from a recent issue:

E. Doedel: The computer-aided bifurcation analysis of predator-prey models
S. Karlin, S. Lessard: On the optimal sex-ratio: A stability analysis based on a characterization for one-locus multiallele viability models
J. M. Mahaffy, C. V. Pao: Models of genetic control by repression with time delays and spatial effects
P. Creegan, R. Lui: Some remarks about the wave speed and traveling wave solutiions of a nonlinear integral operator
H. Aargaard-Hansen, G. F. Yeo: A stochastic discrete generation birth, continuous death population growth model and its approximate solution
F. M. Hoppe: Pólya-like urns and the Ewens' sampling formula
M. Weiss: A note on the rôle of generalized inverse Gaussian distributions of circulatory transit times in pharmacokinetics
R. Dal Passo, P. de Mottoni: Aggregative effects for a reaction-advection equation.

Subscription information and sample copy upon request

Springer-Verlag
Berlin
Heidelberg
New York
Tokyo

Lecture Notes in Biomathematics